PHYSIOLOGY AND TOXICOLOGY OF MALE REPRODUCTION

PHYSIOLOGY AND TOXICOLOGY OF MALE REPRODUCTION

Edited by

JAMES C. LAMB IV

Jellinek, Schwartz, Connolly, & Freshman, Inc.
Washington, D.C.

PAUL M. D. FOSTER

Central Toxicology Laboratory
Imperial Chemical Industries, P.L.C.
Macclesfield, Cheshire
United Kingdom

ACADEMIC PRESS, INC.

Harcourt Brace Jovanovich, Publishers
San Diego New York Berkeley Boston
London Sydney Tokyo Toronto

ACADEMIC PRESS, INC.
1250 Sixth Avenue, San Diego, California 92101

United Kingdom Edition published by
ACADEMIC PRESS INC. (LONDON) LTD.
24–28 Oval Road, London NW1 7DX

Library of Congress Cataloging in Publication Data

Physiology and toxicology of male reproduction.

Includes index.
1. Generative organs, Male. 2. Genetic toxicology.
3. Infertility, Male—Etiology. I. Lamb, James C.,
Date. II. Foster, Paul M. D. [DNLM: 1. Fertility
—drug effects. 2. Genitalia, Male—drug effects.
3. Infertility, Male—diagnosis. 4. Reproduction.
5. Reproduction—drug effects. 6. Toxicology—methods.
WJ 700 P578]
QP 253.P48 1987 616.6'92071 87-17544
ISBN 0–12–434440–2 (alk. paper)

PRINTED IN THE UNITED STATES OF AMERICA

88 89 90 91 9 8 7 6 5 4 3 2 1

Contents

Contributors

Numbers in parentheses indicate the pages on which the authors' contributions begin.

Robert E. Chapin (155), Systemic Toxicology Branch, National Institute of Environmental Health Sciences, Research Triangle Park, North Carolina 27709

E. M. Eddy (35), Gamete Biology Section, Laboratory of Reproductive and Developmental Toxicology, National Institute of Environmental Health Sciences, National Institutes of Health, Research Triangle Park, North Carolina 27709

Paul M. D. Foster (1, 7), Central Toxicology Laboratory, Imperial Chemical Industries plc, Macclesfield, Cheshire SK10 4TJ, United Kingdom

Helen Goeden (179), Department of Pharmacology and Therapeutics, University of Calgary, Calgary, Alberta, Canada

Tim J. B. Gray (225), The British Industrial Biological Research Association, Woodmansterne Road, Carshalton, Surrey SM5 4DS, United Kingdom

James C. Lamb IV (1, 137), Jellinek, Schwartz, Connolly, & Freshman, Inc., 1350 New York Avenue, N.W., Suite 400, Washington, D.C. 20005

Andres Negro-Vilar (103), Reproductive Neuroendocrinology Section, Laboratory for Reproductive and Developmental Toxicology, National Institute of Environmental Health Sciences, National Institutes of Health, Research Triangle Park, North Carolina 27709

Richard M. Sharpe (71), MRC Reproductive Biology Unit, Edinburgh EH3 9EW, United Kingdom

Michael D. Shelby (203), Cellular and Genetic Toxicology Branch, National

Institute of Environmental Health Sciences, Research Triangle Park, North Carolina 27709

Marcelo M. Valenca (103), Reproductive Neuroendocrinology Section, Laboratory for Reproductive and Developmental Toxicology, National Institute of Environmental Health Sciences, National Institutes of Health, Research Triangle Park, North Carolina 27709

Harold Zenick (179), U.S. Environmental Protection Agency, Reproductive Effects Assessment Group (RD689), Office of Health and Environmental Assessment, Washington, DC 20460

Preface

Many recent publications attest to the substantial increase in public and scientific concern regarding the potential of industrial and environmental chemicals to interfere with male reproductive functions. Many symposium volumes and book chapters cover selected areas of this field, but currently no text includes a comprehensive description of both general reproductive biology and specific aspects of reproductive toxicology. Extensive knowledge in both biology and toxicology are necessary for the proper design, conduct, and interpretation of studies in male reproductive toxicity. This book is designed to fill that need by including chapters in general male reproduction as well as chapters on the specifics of male reproductive toxicology.

The present volume is directed primarily toward toxicologists who are not experts in male reproductive biology, and it will lend valuable assistance to graduate students and new investigators entering this expanding field. In addition, it will aid those other biological scientists who work only occasionally in male reproductive toxicology and fertility testing.

The book provides essential background information on the fundamentals of male reproductive biology and reviews recent advances in the physiology of male reproductive function. The application of the fundamental concepts of male reproduction to the evaluation of toxicants is then discussed, including an overview of the basic design and limitations of current routine studies in male reproductive toxicology. Readers are referred to the respective regulatory agencies for specific details of the protocols but are provided substantial background in this book.

To properly perform and interpret routine male reproductive toxicity tests, the investigator must understand the fundamentals of reproductive biology. Similarly, one studying specific mechanisms of toxicant action or target organ responses must be prepared to use state-of-the-art methods and remain familiar with new developments in reproductive biology. Within the past several years there has been a greater acceptance of new methods for the evaluation of target organ and target cell responses in the reproductive system, which provide new information on the reaction of the male to toxicants. These advances, described in this book, include new morphological, biochemical, and endocrinological methods and knowledge, which have improved the sensitivity and predictive value of toxicity evaluations. Finally, selected chapters are devoted to the use of *in vivo* and *in vitro* genotoxicity assays and *in vitro* systems for possible mechanistic studies of male reproductive toxicants.

James C. Lamb IV
Paul M. D. Foster

Chapter 1

Introduction

PAUL M. D. FOSTER

Central Toxicology Laboratory
Imperial Chemical Industries
Alderley Park
Macclesfield, Cheshire, SK10 4TJ United Kingdom

JAMES C. LAMB IV*

Office of Pesticides and Toxic Substances
U. S. Environmental Protection Agency
Washington, D.C. 20460

Toxicology is a rapidly expanding field of scientific investigation. This expansion is the result of various factors, such as the increased production of commodity chemicals and the constant development of new compounds each year, that have the potential to cause adverse health effects. Among the adverse effects, those of greatest concern are chronic toxicity, cancer, reproductive dysfunction, and birth defects. Many of the new chemicals are being evaluated for their ability to cause such effects; moreover, many materials that have been used extensively for a number of years are now being reevaluated and have been shown to produce toxicity, including impairment of reproductive function in animal models.

Normal reproduction is dependent on a series of complex biological interactions. The production of a viable offspring will only result following

*Work not performed as part of official duties, does not necessarily represent the opinions of the U.S. Government.

proper male and female reproductive function, and normal development of the conceptus. This complex sequence of events is increasingly recognized as being susceptible to perturbation by physical or chemical environmental changes. A good deal of attention has been focused on female reproduction and teratogenesis by research toxicologists, regulatory agencies, industry, and the general public. Male reproductive toxicology, however, has recently become a rapidly expanding area of research and testing and public concern. It has not been unusual to exclude women of childbearing age from certain occupational environments, but only recently have employers given serious consideration to potential adverse effects on male reproduction. There is also a heightened awareness of the potential for chemical exposures to cause adverse effects on subsequent generations as a consequence of exposure to the male. Despite trends toward an increase in the numbers of women in the workplace, males are still found in the industrial environment more often than females, and there is an increased awareness and concern by industry for reproductive health of male workers.

The changes in attitude of employers and the public toward worker safety in general, and the realization that male reproduction is sensitive to chemical insult, have come from actual experiences where infertility or other reproductive dysfunction have resulted from occupational chemical exposures. Associated with the increased general awareness of the significance of the problems of male reproductive toxicity is a greater appreciation for the need for better ways to determine potential adverse effects on male reproduction.

Examples of chemicals that have adversely affected human male reproduction include 1,2-dibromo-3-chloropropane (DBCP) and Kepone (chlordecone). Both of these chemicals are testicular toxicants in men, and both chemicals were demonstrated testicular toxicants in animal models prior to inadvertent human exposure (see Chapters 6 and 7). These examples provide important lessons that animal data may aid in hazard identification for male reproductive function.

To go beyond simple hazard identification, one must know more about the biological and metabolic processes that occur in the animal species and how adequately these processes model the human response. Knowledge that chemical absorption, distribution, metabolism, and excretion are the same in humans as in the animal model will allow better understanding of potential hazard, which will eventually be applied, with other factors, in risk assessment. These pharmacokinetic–metabolic parameters are chemical specific, and it follows that one species may be an appropriate model for a particular chemical, but not necessarily for others. An example of chemical-specific differences in species sensitivity is the testicular toxicity of DBCP. DBCP was a testicular toxicant in humans and in rats, but it was not a testicular toxicant, even at relatively high dose levels, in the mouse. Although the mouse may predict other testicular toxicants, it would not have predicted the effects of DBCP.

To date, the rat constitutes the species that is most thoroughly studied for general pharmacology, endocrinology, and male reproductive physiology. It should be appreciated that there are anatomical differences, as well as physiological and biochemical differences, that can affect toxicity. Furthermore, the selection of higher mammals, such as primates, does not necessarily provide a better model for male reproductive toxicology than the rat. For example, the testicular anatomy of most primates is not any more similar than the rat to human testicular anatomy.

Although one can point to specific examples of reproductive toxicants and target organs or cells, one cannot identify the full spectrum of male reproductive toxicants. The ultimate purpose of the male reproductive system is to produce and deliver sperm that are capable of fertilizing the ovum and producing normal offspring, regardless of the species of interest. To properly evaluate a chemical for potential effects on male reproductive function, the entire integrated male reproductive system must be considered. There are numerous examples of chemicals that are known to produce infertility in the male. Their initial targets, however, may be quite different and the ultimate result of the exposure, infertility, may be the result of entirely different toxicological processes. Figure 1 outlines the major organ systems or processes that are essential for normal reproduction and gives selected examples that affect them. The disruption or interruption of any one

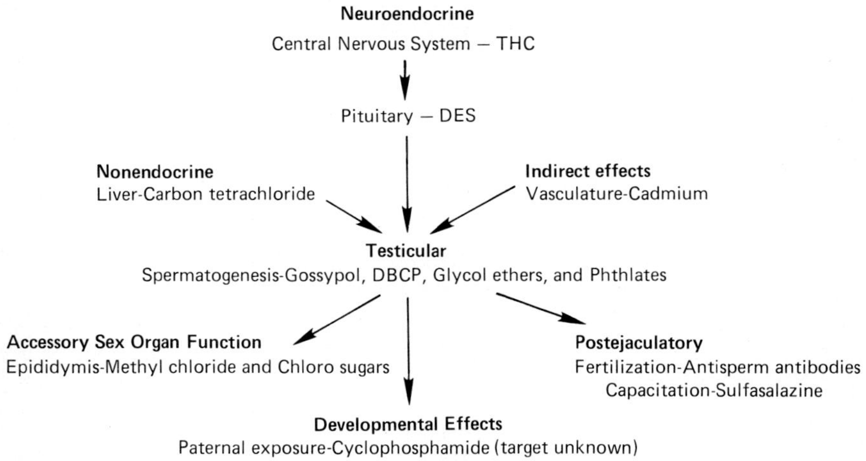

Fig. 1. Illustration of organ systems, organs, or processes that are part of the integrated male reproductive system and examples of toxicants that can affect male reproduction at those sites. THC, tetrahydrocannabinol; DES diethylstilbestrol; DBCP, 1,2-dibromo-3-chloropropane.

of these will lead to infertility or a failure to produce viable offspring. Other chapters in this volume will address the potential target organs or systems in considerable detail.

It is important to reiterate that compounds acting on different processes may lead to the same result in an initial screening for effects on fertility. For example, diethylstilbestrol (DES) will act on the pituitary, gonad, and accessory sex organs, while cadmium can disrupt the vasculature to the testis. These agents both cause effects that will manifest themselves by reduced sperm production and decreased fertility, but subsequent studies must be conducted to determine the ultimate target site for the toxicant. The procedures for properly evaluating male reproductive toxicity include characterization of the toxicity, followed by pharmacokinetic, pharmacodynamic, and metabolism studies from the test species, and dissecting the mode of action of the chemical (Fig. 2). These data are then applied to more accurate hazard identification and, through structure–activity studies and species comparisons of metabolism and pharmacokinetics, hazard avoidance. The development of comprehensive data in such a manner will allow more scientific and relevant risk assessment and risk management.

The present state of the art of risk assessment for adverse effects on reproduction is rather unsophisticated. The toxicity test serves to identify the

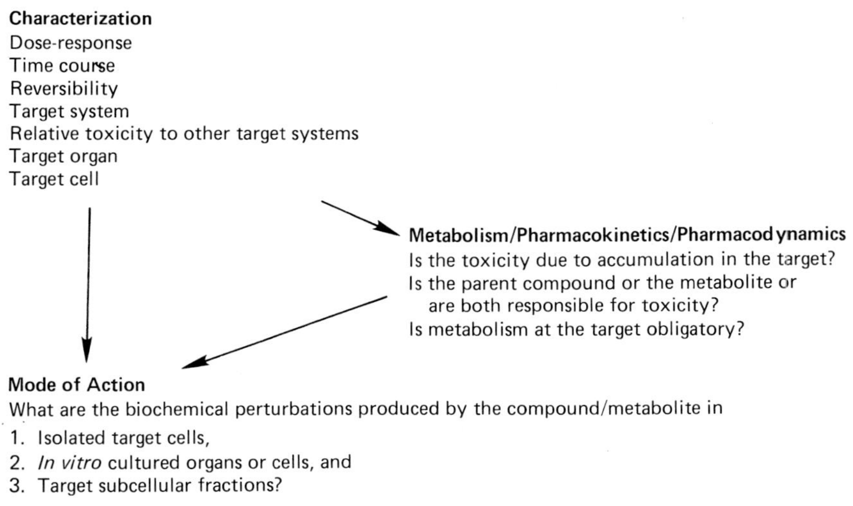

Fig. 2. Rationale for the study of male reproductive toxicants. Male reproductive effects should be studied in an orderly and comprehensive manner. The toxic effect is first characterized in the test animal and compared to human exposure conditions. The movement of the chemical toward the target is studied to determine the proximate toxicant. The mode of action is sorted out within the target cells through various systems.

hazard and should establish the relationship between the dose and the response. In any attempt at the estimation of human risk from animal models, the risk manager must consider the relevance of the biological endpoint that is affected, the route of exposure, and the duration of exposure. A level is identified at which no adverse effect is observed. The effect could be any significant change in the various parameters described throughout this book. Once a no-observable-effect level (NOEL) is found, a safety factor, often 100-fold, is applied to determine the acceptable exposure for humans. The concepts and particulars of risk assessment, and to a limited extent risk management, are described further in Chapters 6 and 8.

There is a good deal of anecdotal information in the literature that alludes to significant reductions in human sperm counts in the general population over recent decades. It is impossible to rely on such reports for any purpose, but they do point to a concern about male reproductive effects. Also, certain investigators point to sperm assessment data in humans as evidence that most men are marginally fertile and, therefore, more susceptible to toxic effects on reproductive function than the animal models that we use. This concept is discussed further in Chapter 8, and may prove important in risk assessment if it holds true. A less fertile strain of rat or mouse, however, would not necessarily provide a better model for human effects, because such a model would probably be more variable and less reliable. As changes in models are discussed in Chapters 5–10, the reader should critically evaluate suggestions in light of that particular experiment's purposes.

In recent years the field of toxicology has come under increased pressure to reduce number of animals used in testing, while at the same time the number of new compounds requiring testing has increased. In attempts to overcome this dilemma there has been general movement toward the use *in vitro* test systems. This has also been true in the field of male reproductive toxicology, with all consequent problems that ensue in attempting to model an integrated biological system (many of these points are discussed in Chapters 5 and 10). In general, our present knowledge dictates that *in vitro* systems be used in attempts to dissect toxicological effects subsequent to an *in vivo* response.

The use of such systems as research tools in the study of the mode of action of male reproductive toxicants and by incorporation into existing and improved test systems should lead to a more rational and scientific approach to hazard assessment and ultimately to the design of safer chemicals. Many of these aspects will be discussed in more detail in chapters concerning testicular organization (Chapters 2, 4, and 7), reproductive toxicity testing systems (Chapter 6), and germ-cell genetic toxicology (Chapter 9).

Crucial to any improvement in this hazard assessment is a thorough understanding of the biology of male reproduction and the type and appropriate utilization of data generated in toxicity testing procedures.

The fundamental reproductive biology of the testis (Chapters 2 and 4), the neuroendocrine system (Chapter 5), the epididymis and accessory sex organs (Chapter 3), and spermatozoal evaluation (Chapter 8) are each discussed in individual chapters to present a thorough overview of this complex, integrated system. Each of the authors has attempted to illustrate the potential of recent studies in the understanding of normal reproductive function by reference to examples of toxicants or, where there exists an opportunity, to incorporate recent findings as indices of perturbation of the relevant system.

The philosophy reflected by this book is that the reproductive toxicologist must be expert not only in toxicology, but also in reproductive biology. Therefore, we have provided chapters in normal testicular function and biochemistry (Chapter 2), accessory sex-organ function (Chapter 3), and endocrinology (Chapters 4 and 5). These chapters give the reader a foundation in normal male reproduction and provide some examples of useful endpoints which can be measured.

Once the foundation in normal function is established, the reader proceeds to chapters that are toxicologically oriented. Chapter 6 describes the toxicological evaluation of the complete reproductive system, followed by chapters on testicular morphology (Chapter 7), sperm assessment (Chapter 8), and germ-cell mutagenesis (Chapter 9). The book is concluded with a description of new *in vitro* methods that are being developed in the area (Chapter 10). Together these chapters could provide the reader with essential knowledge to design, conduct, or evaluate reproductive toxicity studies.

Chapter 2

Testicular Organization and Biochemical Function

PAUL M. D. FOSTER

Central Toxicology Laboratory
Imperial Chemical Industries
Alderley Park
Macclesfield, Cheshire, SK10 4TJ United Kingdom

I. Introduction

The objectives of this chapter are (1) to provide information on testicular structure and compartmentation, (2) to describe the process of spermatogenesis with special regard to its cyclical and synchronous nature, and (3) to discuss the role of the Sertoli cell in providing close control of the process. Aspects of hormonal testicular control both via the pituitary and intragonadally will make reference to the detailed presentations made in Chapters 4 and 5. Reference will be made to examples of toxicants that perturb normal testicular function, and again the reader will be referred to Chapters 6 and 7.

Two prime functions can be ascribed to the testis: first, the generation of mature gametes that will be capable of fertilizing an oocyte to produce viable offspring, and second, the production of testosterone, the major androgen, which maintains spermatogenesis (discussed later) and is secreted into the circulation to initiate, support, and maintain secondary sexual characteristics, for example, the development of the accessory sex organs (see Chapter 3).

A number of components are essential for the initiation and maintenance of spermatogenesis [e.g., vitamin A (Thompson *et al.,* 1964) and zinc (Barney *et al.,* 1969; Lei *et al.,* 1976)], but little is known as to why they are important or what their precise role is in assisting normal testicular function. In general, our biochemical understanding of testicular function is relatively poor compared to other major organs (e.g., liver and muscle).

The testis is known, for instance, to possess unique isozymes of important enzymes in intermediary metabolism [e.g., LDH-C4 (Goldberg, 1977)] and detoxification [e.g., glutathione transferases (Guthenberg *et al.,* 1983; Boyer and Kenny, 1985)], and many of these have activities and substrate requirements different from those in the liver. The significance of these differences is largely unknown, and only recently have studies commenced to understand the reasons for possible specificity and the need for discrete isozymic patterns. Indeed, we still do not understand the precise cellular role for an agent as important as testosterone; although data have been available describing its essential nature for decades, the full picture of the processes regulated by testosterone is incomplete.

Throughout the remainder of this chapter, comments on testicular structure and function will be addressed using the rat as the model. This species has been chosen for a number of reasons: particularly, that the rat has provided the major source of information regarding normal testicular structure, physiology, and biochemistry. In addition, it is the species used to the largest extent in routine toxicology and reproductive toxicology studies and necessarily where the vast majority of male reproductive toxicants have been identified and further investigated. It is probably fair to state that all of the known human male reproductive toxicants are known to be active in the rat, although the converse is certainly not proven.

II. Testicular Compartments

The parenchymal tissue of the testis is enclosed within the tunica albuginica and can be conveniently divided into two functional compartments, the vascularized interstitial tissue and the seminiferous tubules. Spermatogenesis, which will be discussed later, takes place in the tubules, which are long convoluted structures connected at both ends into the rete testis. Sperm are then channeled from the rete via efferent ducts to the epididymis (see Chapter 3).

The interstitial compartment contains Leydig cells, macrophages, and mast cells (see Fawcett *et al.,* 1973; Clark, 1976; Connell and Connell, 1977; Ewing, 1983). The major function of the Leydig cell is the production of testosterone and other steroid hormones to support the inititation and maintenance of spermatogenesis (discussed later). In the rat the volume of the Leydig cells is small. The cells are generally clustered around the blood capillaries contained in sparse connective tissue. In humans, clusters of Leydig cells are found in abundant connective tissue drained by lymphatic vessels (for review, see Fawcett *et al.*, 1973). The relationship of intersitial tissue to the seminiferous tubules is clearly illustrated in the electron micrograph of Fig. 1.

The boundary tissue of the seminferous tubule has a clearly defined layered structure of adventitial cells. The myoepithelial layer of peritubular cells was thought to be concerned only with aiding the passage of released sperm along the tubule; more recent investigations have indicated that their role may be more closely associated with the Sertoli cell and the process of spermatogenesis.

Within the tubules are the somatic Sertoli cells and germ cells at different stages of maturity. The Sertoli cell extends from the periphery to the lumen of the tubule (see Figs. 2 and 3) and fulfills a number of important functions which are discussed later. One of these is the formation of a functional "blood–tubule" barrier by the development of specialized tight junctions between Sertoli cells (see reviews by Dym and Fawcett, 1970; Fawcett, 1975; Setchell and Waites, 1975; Russell, 1978). The inter-Sertoli cell spaces can be divided into the basal and adluminal compartments (see Fig. 3). The basal

Fig. 1. Scanning electron micrograph showing seminiferous tubules and interstitium. (Courtesy of J. R. Foster.)

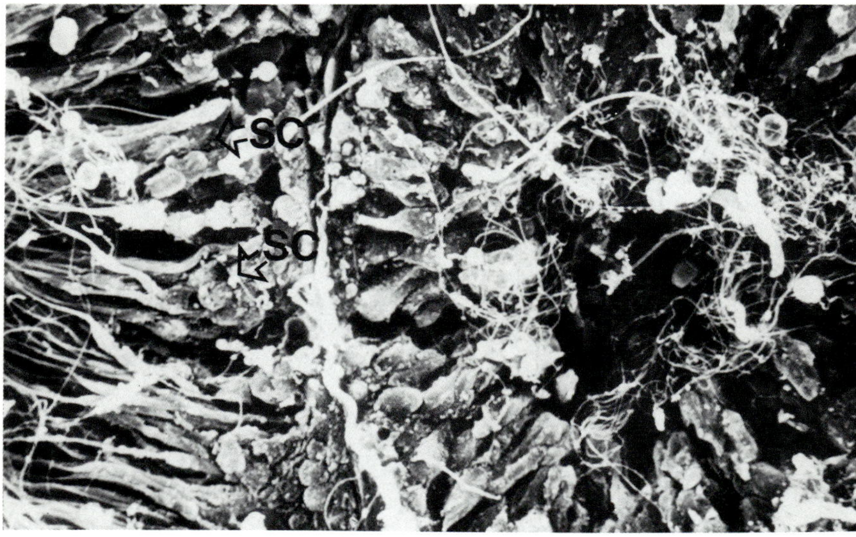

Fig. 2. Scanning electron micrograph of two seminiferous tubules showing arrangement of germ cells about Sertoli cells (SC). (Courtesy of J. R. Foster.)

compartment containing the peritubular cells and spermatogonia would be freely accessible to blood-borne nutrients and toxicants. The adluminal compartment, however, constitutes a "protected" environment and would presumably contain components processed and/or transported by the Sertoli cell. An intermediate compartment also exists whereby preleptotene spermatocytes are "processed" from the basal to the adluminal compartments and further maturation (Russell, 1977).

These functional compartments may have a profound influence on how toxicants may perturb normal testicular function, as they provide both targets for toxicity and variables that must be included in the assessment of toxicity. Important toxicological questions relate to (1) the passage of exogenous compounds from the bloodstream to the Leydig cells and spermatogonia, (2) whether they cross the Sertoli–Sertoli cell barrier (intact and/or modified), and (3) what further effects are produced within the adluminal compartment. Concentration of a potential toxicant in any of these compartments may be a determining factor in the testicular damage that may ensue.

Although tissue distribution studies have been conducted with known testicular toxicants, there is little evidence to suggest that toxicity results from accumulation of the compound in the target organ. Small amounts of label in the whole organ tend to preclude accurate measurements in specific compartments, but due consideration must be given to the fact that even though small amounts of material may reach the testis, toxicity may result from specific accumulation within certain cells or compartments of the organ.

III. Spermatogenesis

The development of the male gamete in the seminiferous tubules is dependent on the action of both the pituitary gonadotrophins (see Chapters 4 and 5) and androgens secreted by Leydig cells (Steinberger, 1971). It is now

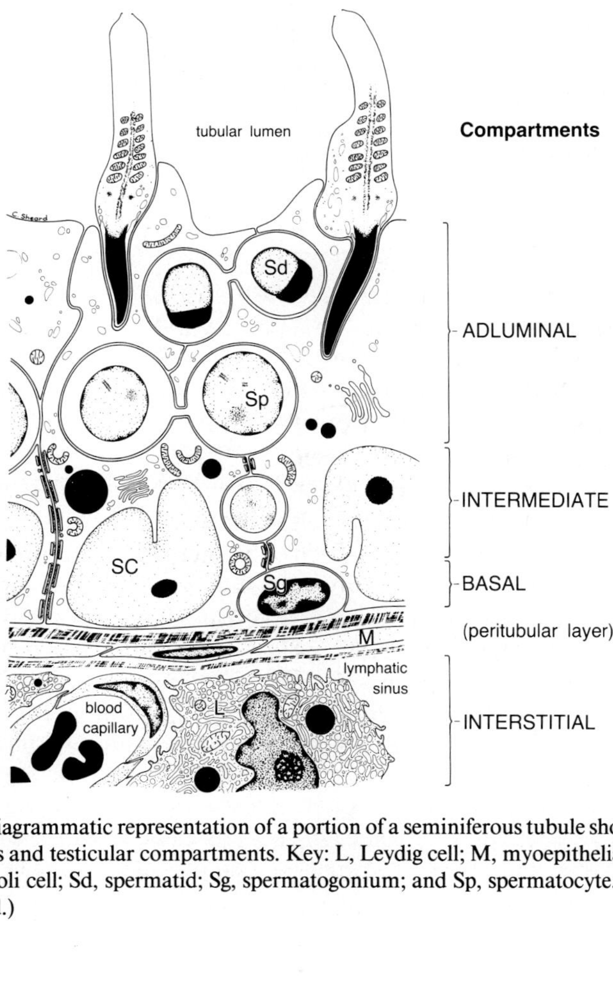

Fig. 3. Diagrammatic representation of a portion of a seminiferous tubule showing cellular arrangements and testicular compartments. Key: L, Leydig cell; M, myoepithelial peritubular cell; SC, Sertoli cell; Sd, spermatid; Sg, spermatogonium; and Sp, spermatocyte. (Courtesy of C. M. Sheard.)

generally accepted that the Sertoli cell is the primary mediator of spermatogenesis, it being the major testicular target for the action of the pituitary gonadotrophin, follicle-stimulating hormone (FSH), and androgens. The precise regulation of the process (see review by Parvinen, 1982) will be discussed in more detail in later chapters.

Spermatogenesis is a dynamic process whereby undifferentiated stem cells (spermatogonia) develop into highly specialized spermatozoa. The process can be divided into four distinct phases:

1. the proliferation and renewal of undifferentiated spermatogonia;
2. the differentiation of spermatogonia;
3. the process of meiosis; and
4. spermatid development (spermiogenesis) and release of spermatozoa (spermiation).

During the first phase, the undifferentiated type A spermatogonia (found in the basal compartment of the tubule outside of the "blood–tubule" barrier) divide mitotically to yield one of two products. These can be two daughter cells, which separate to become new type A stem cells, or two cells that remain joined by a cytoplasmic bridge and are committed to differentiate (second phase). The factors controlling stem-cell behavior are not well understood but are believed not to be hormonal (e.g., proliferation will occur in hypophysectomized animals) and may be due to factors produced by the Sertoli cell (Bellvé and Feig, 1984). Differentiated spermatogonia destined to proceed to meiosis (third phase) undergo six mitotic divisions through intermediate and type B spermatogonia. After the last mitosis, preleptotene or resting spermatocytes are formed. As these cells develop into leptotene spermatocytes they "pass through" the blood–tubule barrier. This event involves the adjacent Sertoli cells interdigitating portions of their cytoplasm peripheral to the spermatocytes, where they meet and form new tight junctions, while the existing junctions dissolve; thus the leptotene spermatocytes enter the adluminal compartment and continue meiotic prophase. Leptotene is followed by zygotene, during which homologous chromosomes are paired. Cells with completely paired chromosomes are termed *pachytene spermatocytes.* Pachytene is the longest step of meiotic prophase, followed by a short diplotene step, during which the chromosomes partially split and finally the nucleus undergoes the first meiotic division to yield secondary spermatocytes. These short-lived cells divide again without replicating their DNA to form the haploid spermatids. During spermiogenesis (fourth phase) these early, round spermatids develop an acrosome (an organelle containing lysosomal enzymes that will enable the sperm to penetrate the zona pellucida of the oocyte for fertilization) from the Golgi complex, elongate, condense their nuclei, and develop a tail.

During spermiogenesis specializations appear between the spermatids and the Sertoli cell that are thought to help ensure smooth transfer of germ cells toward the lumen of the seminiferous tubule (for review see Russell, 1980). At spermiation, testicular spermatozoa are released into the tubular lumen. At this point these cells both are nonmotile and lack the ability to fertilize an ovum. Most of the spermatid cytoplasm and organelles remain to form a "residual body," which is immediately phagocytosed and then processed by the Sertoli cell.

The whole process is depicted diagrammatically in Fig. 4, such that the time taken for a differentiated spermatogonium to become a testicular spermatozoon is approximately 8 weeks in the rat. In this species a second wave of spermatogonial differentiation is triggered approximately every 13 days, which coincides with the duration of the spermatogenic cycle. Figure 5 illustrates that if a longitudinal section of a tubule is cut to cover one spermatogenic cycle, it will contain representatives from $4\frac{1}{2}$ successive spermatogenic "waves." The consequence of this is that a cross section of the testis (Fig. 6) will contain various generations of germ cells represented at specific stages of their development. These cellular associations are very specific,

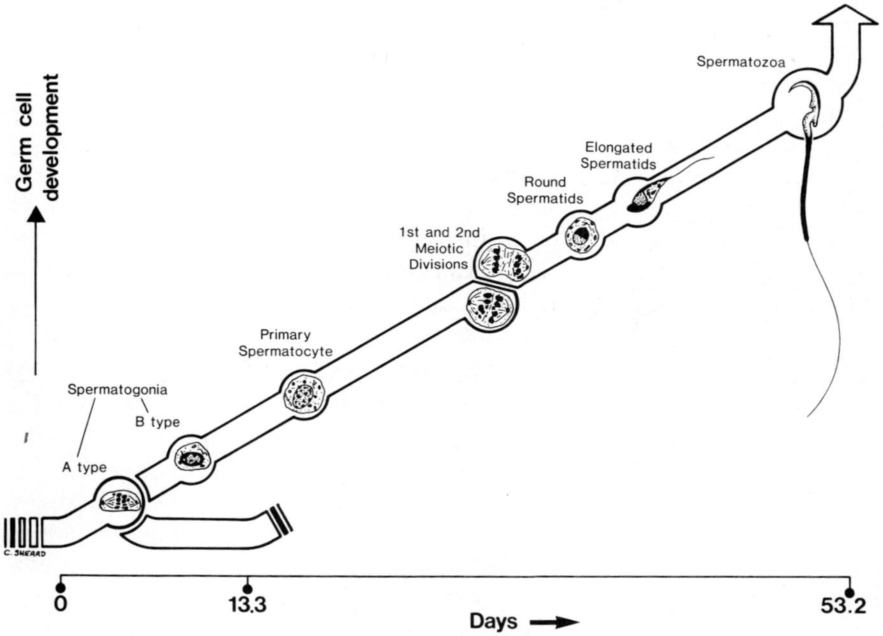

Fig. 4. Diagrammatic representation of spermatogenesis in the rat. (Courtesy of C. M. Sheard.)

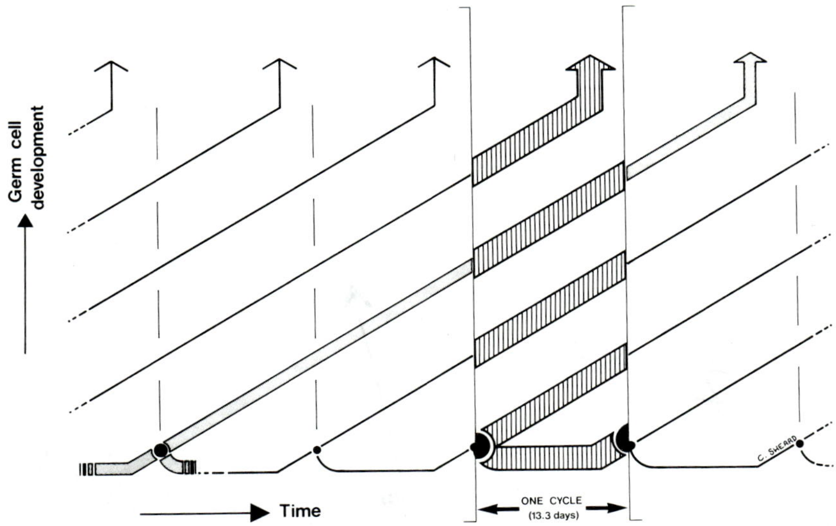

Fig. 5. The cyclicity of the seminiferous epithelium of the rat. The process whereby successive spermatogenic "waves" (see Fig. 4) are interrelated is illustrated. A transverse section of a tubule covering one cycle of the epithelium will contain cell representatives at different stages of maturity from successive "waves." (Courtesy of C. M. Sheard.)

with 14 different stages identified in the rat (Leblond and Clermont, 1952) (see Fig. 7). Each stage includes one or two generations of spermatogonia, spermatocytes, and spermatids. This classification is based on detailed cytological analysis of sections of seminiferous tubules following periodic acid–Schiff hematoxylin staining, with each stage having a characteristic duration. The importance of identifying specific stages of spermatogenesis is related to the cyclic nature of, and the biochemical processes that occur during, the maturation of the germinal epithelium.

Concomitant with the synchronicity of the germ cell production, the Sertoli cell is also recognized as exhibiting cyclic activity in its responses. For example, androgen-binding protein secretion and plasminogen activator concentration and secretion all have peak levels that occur at stages VII and VIII of the cycle in the rat (see review by Parvinen, 1982). Quite interestingly, the Leydig cells of the interstitium (responsible for testosterone production) have also been shown to become closely associated with tubules at these same stages of the cycle (Bergh, 1982, 1985). For more detailed histological information the reader is referred to Chapter 7.

Many of these findings and observations are particularly pertinent to the field of male reproductive toxicology. The introductory chapter has indicated that a large number of theoretical target processes exist whereby chemi-

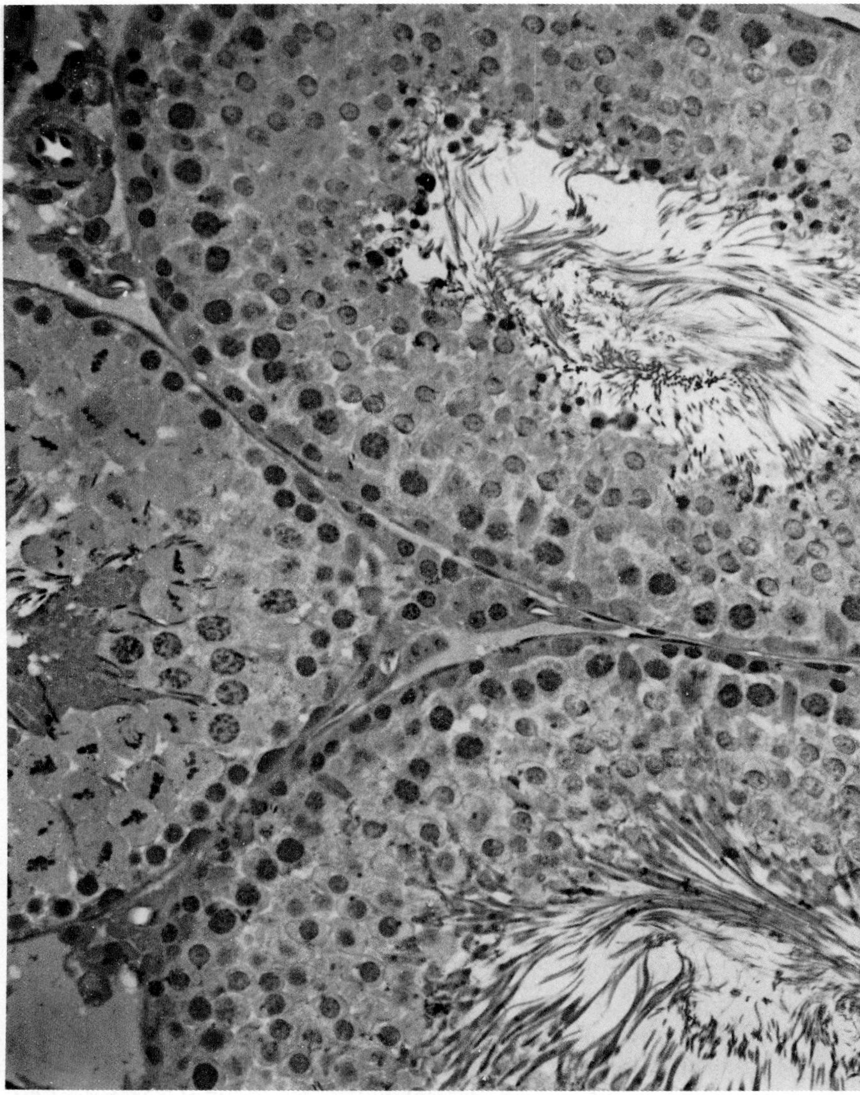

Fig. 6. Cross section of rat testis showing seminiferous tubules with different cellular associations.

cals can alter severely male reproductive function. A survey of the literature of male reproductive toxicants would, however, indicate that the process of spermatogenesis is the one affected by the overwhelming majority of chemicals so far studied. This observation has been based primarily on organ weight changes coupled with testicular morphology in acute or subchronic

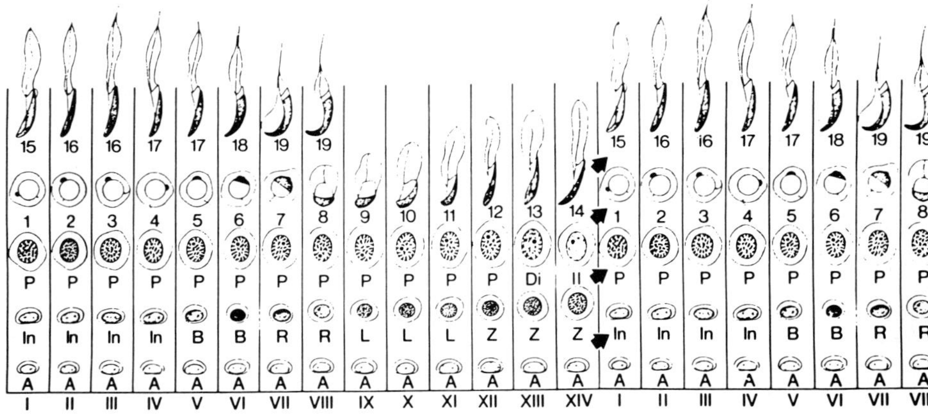

Fig. 7. Diagrammatic representation of the 14 stages of spermatogenesis of rat testis (after Leblond and Clermont, 1952). Each stage (Roman numerals) contains a different cellular association of spermatogonia (A, In, or B), spermatocytes (R, resting/preleptotene; L, leptotene; Z, zygotene; P, pachytene; DI, diplotene; II, secondary), and spermatids (arabic numerals 1–19).

studies. Typically, these studies report an end-stage lesion, rather than initial events from which target-cell toxicity and possible mechanism can be inferred. However, where reasonable time-course studies have been conducted, they would support the notion that spermatogenesis is the end point most sensitive to the action of toxicants. In the few instances where a toxicant has been shown to act indirectly, such as on the pituitary or testicular vasculature, a characteristic lesion in the seminiferous epithelium is also observed. Some examples of these effects are discussed in more detail in Chapter 7.

Within the process of spermatogenesis, each of its cellular components may be a "target" for the action of a compound. For each different testicular cell type, one can find in the literature a compound that appears to target that cell type directly (Table I). Moreover, data have more recently appeared to indicate that toxicants may affect the seminiferous epithelium in a stage-specific manner. This may ultimately provide insights into biochemical processes being disturbed by these compounds as our knowledge of the cyclical nature of the epithelium increases and vice versa. For instance, 2-methoxyethanol, a water-soluble organic solvent, has been shown to produce specific damage to pachytene, diplotene, and secondary spermatocytes in stages XIII, XIV, and I at doses of 100 mg/kg within 24 h of a single oral administration (Foster *et al.,* 1983; Creasy and Foster, 1984; Chapin *et al.,* 1984a). As another example, many hormonally active compounds have been shown to particularly affect pachytene spermatocytes and spermatids at stage VII

TABLE I

Target Cells in the Process of Spermatogenesis

Testicular target	Toxicant	Reference
Leydig cells	Ethanol	Van Thiel *et al.* (1983)
	Ethanedimethane sulfonate	Rommerts *et al.* (1985)
Sertoli cells	Phthalate esters	Foster *et al.* (1982)
	2,5-Hexanedione	Chapin *et al.* (1983)
Spermatogonia	Busulphan	Jackson *et al.* (1962)
Spermatocytes	2-Methoxyethanol	Foster *et al.* (1983)
		Chapin *et al.* (1984)
Spermatids	Ethyl methane sulfonate	Jackson *et al.* (1961)
	Methyl chloride	Chapin *et al.* (1984b)

(Russell *et al.,* 1981), which is presumably related to the high testosterone requirement at this stage (see above). Therefore, for a compound of unknown activity the observation of specific damage in stage VII would give a strong indication that testosterone production or utilization had been impaired.

Damage to different cell types can theoretically have profound differences in severity of response and recovery from a testicular lesion and effects on fertility following toxic insult. A lesion to spermatocytes or spermatids would tend to result in decreased fertility that is fairly rapid in onset (allowing for a 13-day epididymal transit time), but that would be expected to be reversible as daughter cells replace the affected population. Widespread damage to spermatogonia would be likely to produce the most severe effects, with decreased fertility taking more time to appear (6 – 8 weeks in the rat) but being more protracted (i.e., with no stem-cell population), and succeeding populations of germ cells and ultimately spermatozoa would not be produced. Damage to Sertoli cell function, with inadequate physical and biochemical support to the germ cells, would result in immediate effects on spermatogenesis. If the integrity of the epithelium is not maintained, infertility would be lengthy and possibly permanent if widespread Sertoli cell death ensued, since these cells normally do not divide after puberty.

Although the experimental data base illustrating these effects on fertility is not large, a comparison of the effects of 2-methoxyethanol [a spermatocyte toxicant (Chapin *et al.,* 1985)] and some work from our own laboratory on 1,3-dinitrobenzene [DNB, a Sertoli cell toxicant (Foster *et al.,* 1986)] will exemplify the different reproductive outcomes following toxicant exposure. Using fertility as a gross end point of reproductive function, Fig. 8 shows that 2-methoxyethanol produces significant effects on fertility at week 5 in a serial breeding study (i.e., the sperm affected were late-stage spermatocytes at the time of treatment 5 weeks earlier). This is consistent with the histological

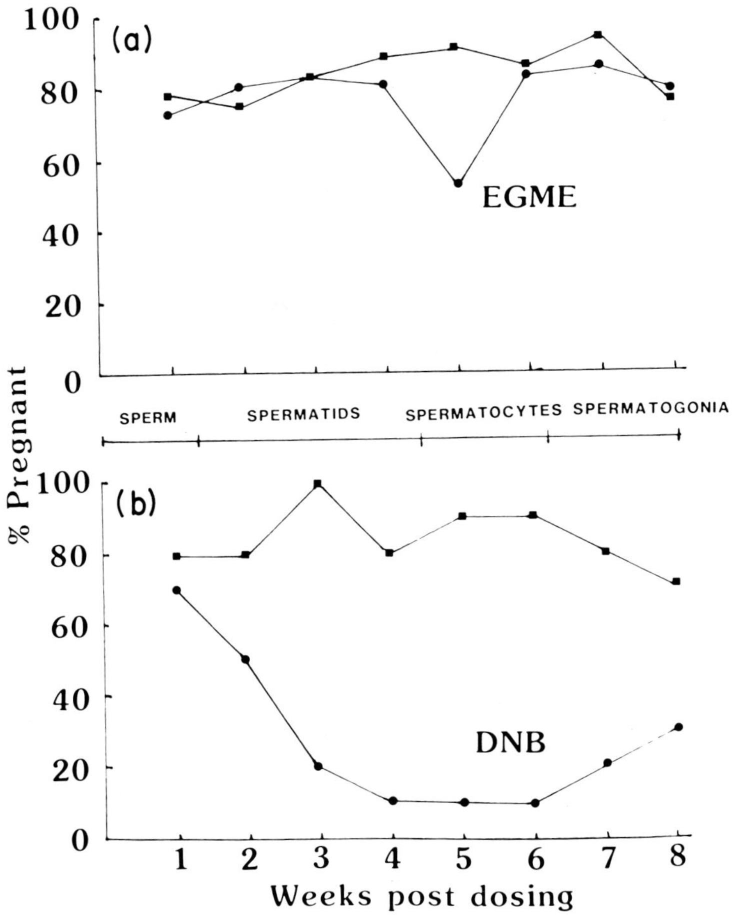

Fig. 8. Effects of (a) 2-methoxyethanol (ethylene glycol monomethyl ether, EGME, 100 mg/kg day for 5 days) and (b) 1,3-dinitrobenzene (DNB; 10 mg/kg day for 5 days) on fertility in a serial breeding study: ■, control; ●, treated values [Part (a) from Chapin *et al.* (1985), with permission]. Depressions in fertility are related to the specific types of spermatogenic cells affected.

pattern of damage observed in the testis (Fig. 9). In the case of DNB, apart from epididymal sperm (week 1), a diminution of fertility occurs over the whole range of spermatogenic stages from week 2 (late spermatids) to week 8 (type A spermatogonia) in a manner consistent with a primary effect on the Sertoli cell (see Fig. 10). Thus, 2-methoxyethanol produces reversible, specific dificits in fertility, while no recovery of fertility is seen with DNB after one spermatogenic wave of 8 weeks.

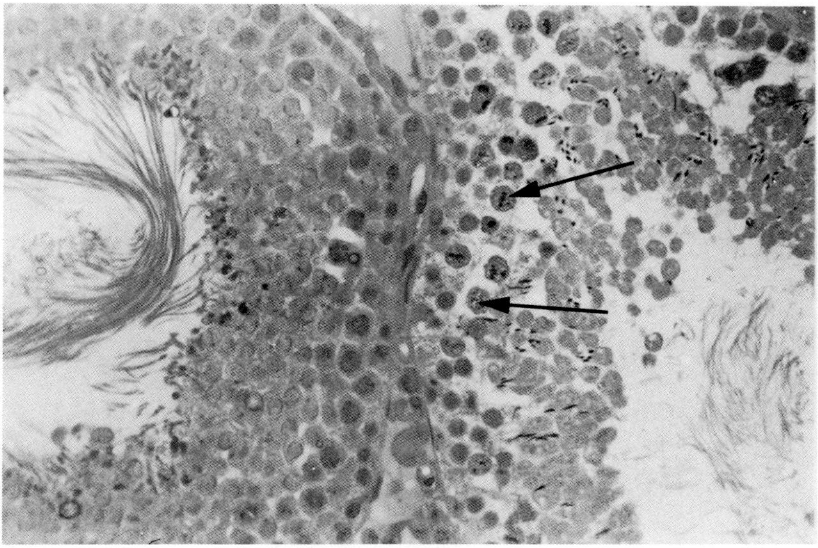

Fig. 9. Effect of EGME (100 mg/kg) on rat testis 24 h after a single oral dose. Note damaged spermatocytes (arrows) in tubule on right adjacent to a normal tubule.

IV. The Key Role of the Sertoli Cell

The reader will have gathered up to this point that the Sertoli cell can be ascribed a pivotal position in the normal function of the testis, and a more detailed inspection of this cell's function and structure is warranted.

A. Sertoli Cell Structure

The Sertoli cells are the only somatic cells within the seminiferous tubules. They were first described by Sertoli in 1865 as cells whose cytoplasm extended from the peripheral basal lamina of the tubule to the lumen. The Sertoli cells have numerous cytoplasmic processes that completely envelop the developing germ cells to provide physical structure to the germinal epithelium. The close physical contact between the Sertoli cell and other germ cells has led to the concept that the Sertoli cells provide a "nurse" function within the germinal epithelium. This chapter will only allow a brief resume of Sertoli cell structure, and the reader is directed for more detailed information to the reviews by Fawcett (1975) and Russell (1980).

The nuclei of the Sertoli cells tend to lie toward the basal area of the tubule, parallel to the spermatogonia, although some changes in shape and position do occur through the spermatogenic cycle. Electron-microscopic

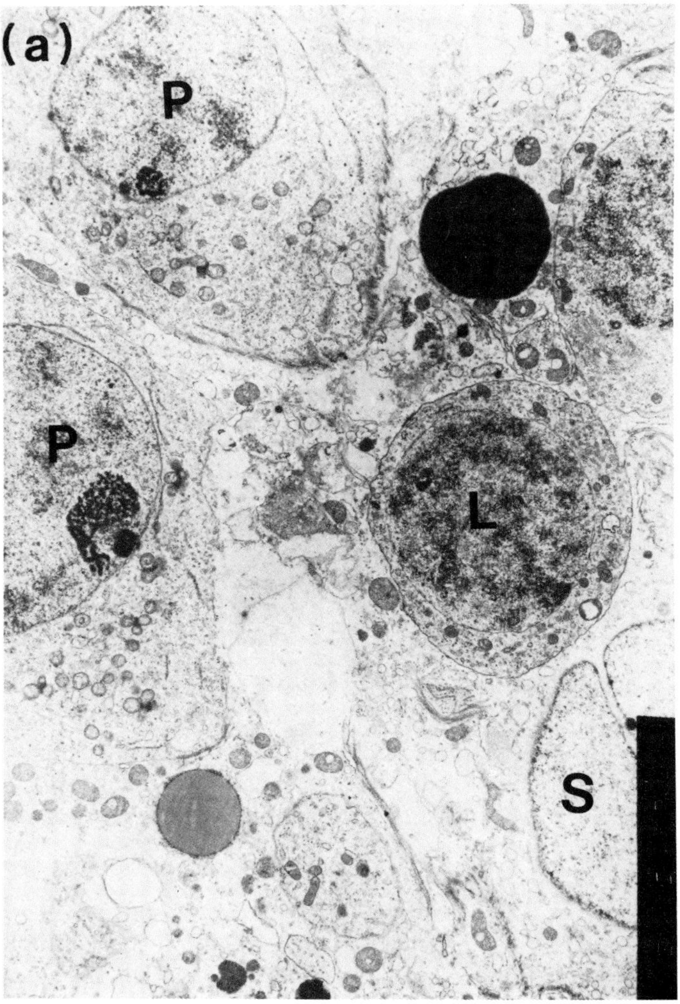

Fig. 10. (a) Electron micrograph of a seminiferous tubule from a 1,3-dinitrobenzene–treated rat (25 mg/kg) after 24 h, showing prominent vacuolation of Sertoli cell cytoplasm. (b) Micrograph from corresponding control animal: S, L, and P refer to Sertoli cell nucleus, leptotene, and pachytene spermatocytes, respectively.

examination has shown the nucleus to be irregular and indented, with a tripartite nucleolus. The perinuclear and basal regions of the cell contain a rich supply of organelles, including endoplasmic reticulum, a specialized Golgi system (consisting of a number of elements believed to be involved with the many secretory products known to be released by Sertoli cells into

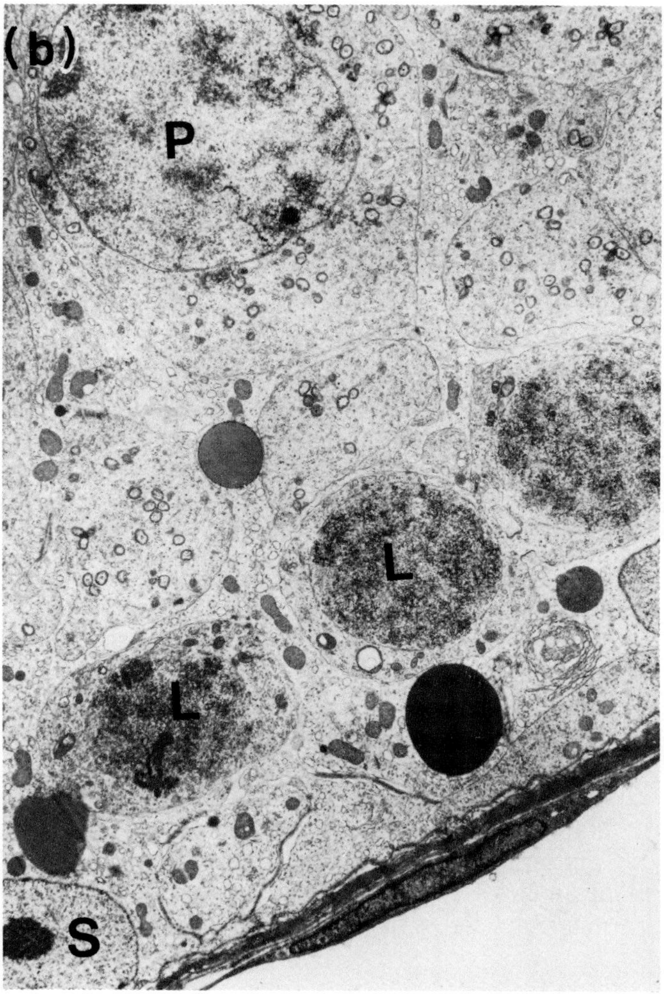

Fig. 10. (*Continued*)

the tubular lumen), and lysosomal structures that are assumed to aid in the phagocytic function of the Sertoli cell, as it ingests degenerate germ cells and the residual body. The residual body is presumed to be responsible for the variable amounts of lipid located in Sertoli cells during specific stages of the cycle. It is thought that this lipid material, together with the presence of all the necessary organelles associated with steroid-secreting cells, may enable the Sertoli cell to produce its own locally acting hormones at certain specific stages of the spermatogenic cycle and thereby to influence germ-cell development.

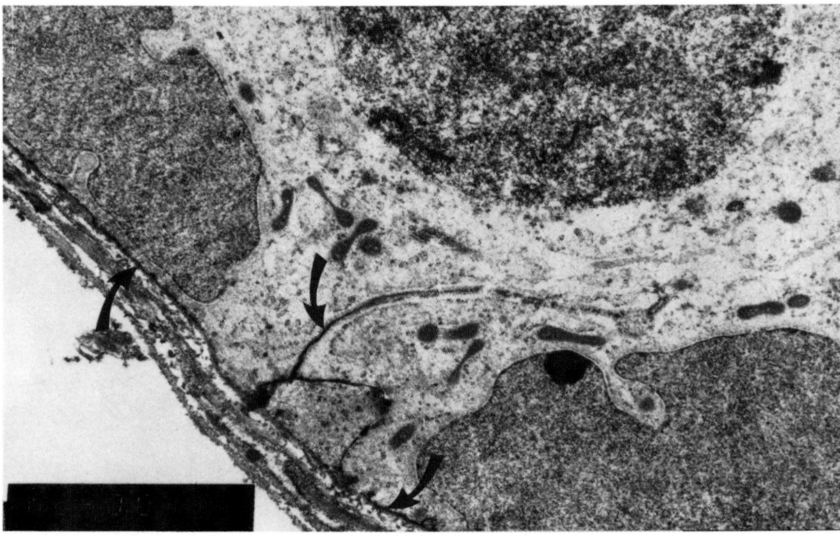

Fig. 11. Electron micrograph of a seminiferous tubule after lanthanum perfusion to indicate the blood–tubule barrier (arrows).

As was mentioned earlier, Sertoli cells produce junctional complexes between each other that form the blood–tubule barrier (see reviews by Dym and Fawcett, 1970; Setchell and Waites, 1975), and the barrier may be visualized using electron-dense substances such as lanthanum (Fig. 11). The formation of this permeability barrier occurs at around 15 days of age in the rat, when the Sertoli cells cease division and the first spermatocytes are appearing in the tubules. Reference has already been made to the passage of leptotene spermatocytes across this barrier (Russell, 1977).

B. Sertoli Cell Function

The steroidal and hormonal control of Sertoli cells is discussed in Chapter 4. Sertoli cells are known to have specific receptors for FSH and will take up testosterone secreted by Leydig cells. The Sertoli cell is now known to be responsible for secretion of fluid into the tubular lumen (Setchell, 1969) to enable the passage of sperm, together with the production and secretion of a number of specific proteins. These include androgen-binding protein (ABP; Fritz *et al.,* 1976), which will bind and transport testosterone and its more androgenic metabolite dihydrotestosterone (also a Sertoli cell product), transferrin (Skinner and Griswold, 1980), and a plasminogen activator (Lacroix *et al.,* 1977). Although the precise role of these proteins is unclear (they are under FSH control, presumably through a cAMP cascade), they have been used extensively as markers of normal Sertoli cell function in the fields of physiology and toxicology.

The Sertoli cells are also thought to secrete factors involved in gonadotrophin feedback (inhibin) and in paracrine control of the Leydig cells (see Chapter 4 for details).

A number of compounds are known to have the Sertoli cell as their initial target site in the testis. Examples of Sertoli cell toxicants include certain phthalate esters used as plasticizers in the manufacture of polyvinyl chloride (PVC) (Foster *et al.,* 1982; Creasy *et al.,* 1983), 2,5-hexanedione, the metabolite of the solvent *n*-hexane (Chapin *et al.,* 1983), and 1,3-dinitrobenzene, a widely used organic intermediate (see above; also see Blackburn *et al.,* 1985; Foster *et al.,* 1986). These compounds have been shown to produce profound effects on Sertoli cell morphology and then on germ-cell development; in the case of phthalate esters, deficits in Sertoli cell function (in addition to nonmaintenance of the germinal epithelium) have been measured *in vivo* (Gray and Gangolli, 1986) and *in vitro* (see Chapter 10), even though the blood–testis barrier, as visualized by lanthanum, remained intact (Fig. 12). Sertoli cell damage followed by tubular disruption can also have effects on the Leydig cell population, usually eliciting a hyperplastic effect (see review by De Kretser and Kerr, 1983), presumably as a compensatory response.

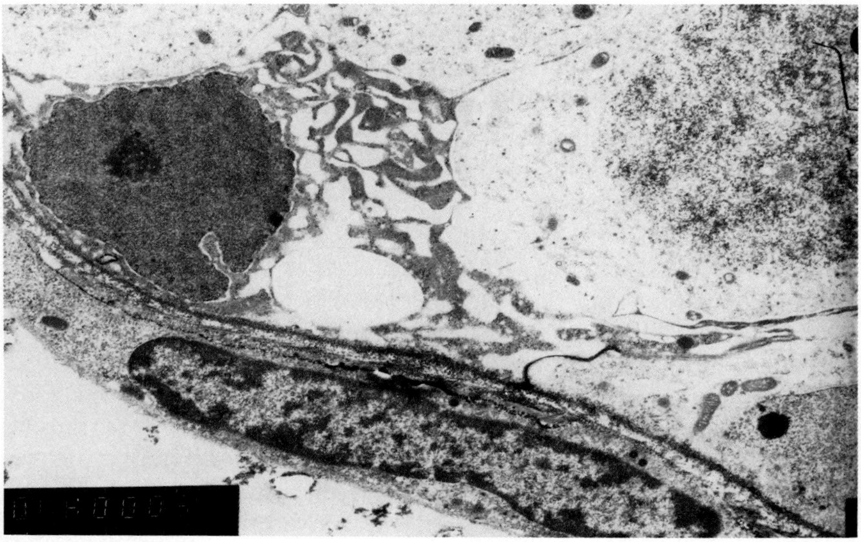

Fig. 12. Electron micrograph as Fig. 11 from rat treated 6 h previously with 2 g/kg of di-*n*-pentyl phthalate. Note severe damage to the Sertoli cell but an intact blood–tubule barrier.

C. *Sertoli–Germ Cell Interactions*

The morphological interactions between Sertoli and germ cells have been described above (see Figs. 2 and 3; also Russell, 1980). The detailed study of Sertoli cell function and its interactions with germ cells, and the attempts to understand the mechanism of action of toxicants have necessitated the development of suitable *in vitro* systems to simply biochemical investigations (see detailed discussion in Chapter 10). In general, our notions regarding the "nurse" function of the Sertoli cell have been confirmed. Although germ cells can be effectively purified by a variety of techniques, they cannot be maintained presently in excess of 24 h. However, when cocultured in the presence of Sertoli cells, they may be maintained satisfactorily for 1–2 weeks and indeed will enhance normal Sertoli cell function (Galdieri *et al.,* 1983, 1984). Sertoli cells can be effectively cultured for similar periods. Recent work (Skinner and Fritz, 1985) has also indicated that coculture of Sertoli cells with peritubular cells will also enhance Sertoli cell function (as estimated by ABP production), posing new questions as to the physiological role of these cells *in vivo.*

Perhaps some of the more interesting aspects of Sertoli–germ cell interactions have focused on the metabolic cooperation between Sertoli cells and spermatocytes and/or spermatids. Some of these are noted in Fig. 13 from recent work regarding carbohydrate and amino acid intermediary metabolism in the testis.

Sertoli cells have been shown to have a high glycolytic rate and secrete lactate and pyruvate, both of which can be stimulated by FSH (Jutte *et al.,* 1983). Additionally, studies on isolated germ cells have indicated that these cells cannot maintain their ATP levels if supplied with glucose, but can do so on addition of lactate and to a certain extent pyruvate (Jutte *et al.,* 1981; Mita and Hall, 1982; Grootegoed *et al.,* 1984). The potential therefore exists for a Sertoli cell product to be a substrate for normal germ-cell function. Interestingly, these same germ cells possess the germ-cell-specific isoenzyme of lactate dehydrogenase (LDH-C_4; Meistrich *et al.,* 1977), which has a greater affinity for lactate than the other isomeric forms of LDH, and hence to utilize lactate for energy production. A similar type of metabolic cooperation may exist in the polyol pathway (see Fig. 13). Ludvigson *et al.* (1982) have elegantly shown using immunocytochemistry that the enzyme aldose reductase is specifically located in Sertoli cells. This enzyme is responsible for the reduction of aldoses to polyols (e.g., glucose to sorbitol); moreover, sorbitol dehydrogenase has been used as a marker of germ-cell number (Hodgen and Sherins, 1973), this enzyme converting sorbitol (and other polyols) to fructose (and other ketoses). The final example concerns the metabolism of branched-chain amino acids, where the aminotransferase needed for the conversion to an α-ketoacid (e.g., valine to α-ketoisovalerate)

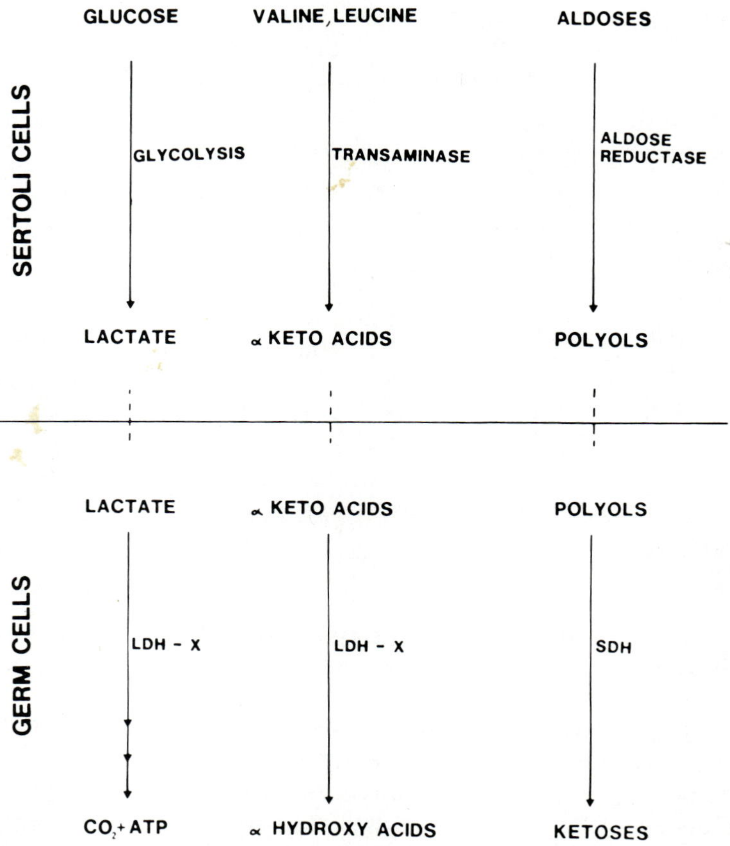

Fig. 13. Sertoli–germ cell biochemical interactions.

has been located in Sertoli cells (Grootegoed *et al.,* 1985) but not in germ cells. Such α-ketoacids (α-ketovalerate, α-ketoisocaproate, etc.) are specific substrates for LDH-C_4 (see Fig. 14) but not for other LDH isozymes, raising the possibility that this enzyme is involved in the cytosolic shuttling of NAD/H in germ cells to facilitate lactate utilization for energy.

These are just a few examples of the many possible metabolic interactions between Sertoli cells and germ cells, which toxicologists are now investigating for potential use as indices of Sertoli–germ cell function and viability. For example, lactate production by Sertoli cells has been shown to be altered by the administration of certain phthalate monoesters (Chapin *et al.,* 1986; Williams and Foster, 1987), 1,3-dinitrobenzene (Williams and Foster, 1987), and dibromochloropropane (Miller *et al.,* 1985).

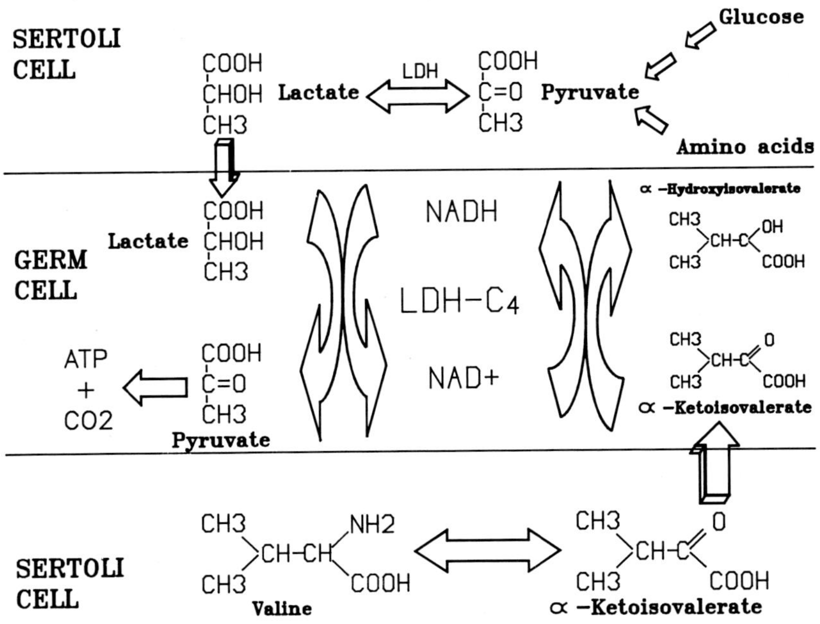

Fig. 14. Relationship between branched-chain amino acid and lactate metabolism in Sertoli and germ cells.

Culture systems are being increasingly used to investigate the mode of action of toxicants once testicular toxicity has been observed *in vivo.* Here again, the usual provisos for the correct use of *in vitro* systems apply (see also Chapters 1 and 10). In common with many other *in vitro* systems, it was believed that Sertoli and Sertoli–germ cell cocultures had little ability to undertake xenobiotic metabolism. Gray and Beamand (1984) have shown that the administration of the *in vivo* toxicants di-2-(ethylhexyl)phthalate or 2-methoxyethanol did not produce effects *in vitro.* However, the metabolites of these compounds—namely, mono-2-(ethylhexyl) phthalate and methoxyacetic acid—are active at producing effects analogous to those seen *in vivo,* indicating that the esterase and alcohol dehydrogenase activities in the cultures may be low. However, such cultures do possess the cytochrome P-450 hemoprotein (and probably, therefore, some mixed-function oxidase activity), which would confirm the steroidogenic potential of the Sertoli cell, since this enzyme complex is intimately involved in the metabolism of cholesterol to testosterone (discussed later). In the Sertoli cells of young animals at least, cytochrome P-450 dependent aromatase activity has been measured (by the conversion of testosterone to 17β-estradiol) and the estradiol produced is thought to inhibit Leydig cell testosterone production.

We have also recently been able to show (Foster *et al.,* 1987) that both

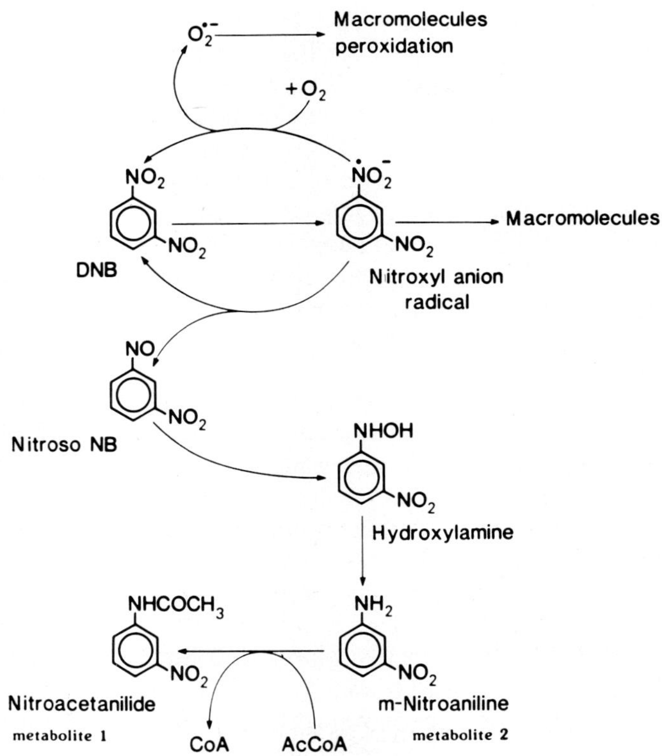

Fig. 15. Metabolism of 1,3-dinitrobenzene in Sertoli cells, showing metabolites identified and proposed route, through possible reactive intermediates, which may result in toxicity. [Reprinted with permission from Foster *et al.*, (1987). Copyright (1987), Pergammon Press, Ltd.]

Sertoli cell and Sertoli–germ cell cocultures will produce metabolites upon incubation with [^{14}C]-1,3-dinitrobenzene (Fig. 15), indicating that xenobiotic metabolism can take place in these cells *in vitro,* and further that metabolism at the target cell may be responsible for specific toxicity seen in the testis.

V. The Leydig Cell

In the rat, Leydig cells are relatively small (~ 10 μm diameter) and are found clustered around the blood vessels of the interstitium (see reviews by Clark, 1976; Connell and Connell, 1977), although their position and association with seminiferous tubules does vary during the spermatogenic cycle (Bergh, 1982, 1985). Bergh (1985) has also indicated that the Leydig cells become particularly associated with tubules where testosterone requirements are highest (stages VII and VIII). In common with many other steroid secretory cells, the Leydig cells possess mitochondria with tubular cristae and have extensive smooth endoplasmic reticulum (SER), although the extent varies between species (Zirkin *et al.,* 1980). A clear, direct relationship exists between the synthesis and secretion of testosterone (the major testicular androgen) and the amount of Leydig cell SER. As has been mentioned previously, testosterone production appears to be the prime function of the Leydig cell. Leydig cell function, it now appears, is controlled by a multiplicity of factors, which include particularly luteinizing hormone (LH) from the pituitary and estrogens and peptides from the seminferous tubules (see Chapter 4).

Since testosterone has such an important role, its production in the testis should be considered in a little more detail. Cholesterol is an obligatory intermediate in the synthesis of testosterone and is supplied to the Leydig cell either by receptor-mediated endocytosis of lipoprotein or by *de novo* synthesis from acetate. Cholesterol esters are probably stored in the numerous lipid droplets frequently observed in Leydig cells. Cholesterol is metabolized to pregnenolone by a macromolecular complex termed the cholesterol side-chain cleavage enzyme (see reviews by Eik-Nes, 1975; Ewing, 1983). The complex is located in the mitochondrion and contains a flavoprotein and cytochrome P-450, requiring O_2 and NADPH as cofactors. The reaction effects the removal of the six-carbon side chain (in the form of isocaproaldehyde) to leave pregnenolone, probably via hydroxylation at the 20 and 22 positions. This reaction is the rate-limiting step for testicular steroid synthesis and is under LH control (Hall, 1966).

The remainder of testicular testosterone production then occurs outside of the mitochondrion. Pregnenolone is converted to testosterone by enzymes located on the SER of the Leydig cell by reactions including 17α-hydroxylase, 17,20-lyase (both cytochrome P-450–dependent), 17β-dehydrogenase and 3β-dehydrogenase, and Δ^{4-5}ketosteroid isomerase. From pregnenolone, two routes are thought to operate, termed Δ^4 and Δ^5 (Fig. 16), although depending on species, up to four routes are thought to be likely (Ewing, 1983; Chubb and Ewing, 1979).

All of these intermediates have been isolated from testicular tissue and it is known that dihydroepiandrosterone (DHEA), androstenedione, and dihydrotestosterone are secreted into the bloodstream. However, they contribute only a small percentage of the effective androgen concentration in the blood.

A number of compounds are now recognized as having the Leydig cell as their testicular target. Much of this information has been derived from work on isolated cells (see Chapter 10) combined with testicular perfusion and *in vivo* studies. A Leydig cell effect is usually manifest in decreased testosterone/steroid production, either basally or in response to LH challenge. An-

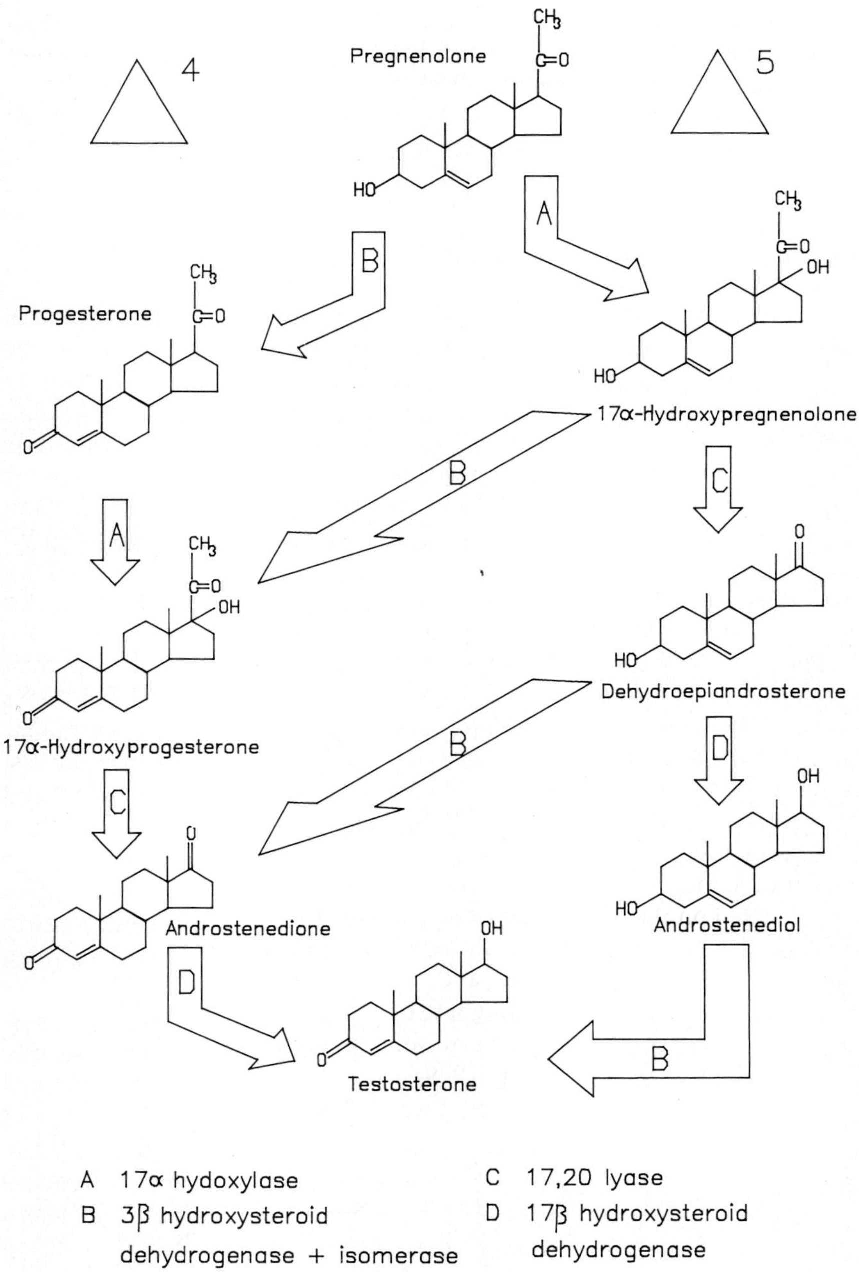

Fig. 16. The metabolism of pregnenalone to testosterone by the Δ^4 and Δ^5 routes: A, 17α-hydroxylase; B, 3β-hydroxysteroid dehydrogenase and isomerase; C, 17,20-lyase; and D, 17β-hydroxysteroid dehydrogenase.

tiandrogens (Russell *et al.,* 1981) and cytochrome P-450 inhibitors (e.g., ketoconazole; Sikka *et al.,* 1985) can be considered as having specific mechanisms, while cadmium (Maines, 1984), ethane dimethane sulfonate (Rommerts *et al.,* 1985), and ethanol and its metabolite, acetaldehyde (Van Thiel *et al.,* 1983), are examples of more nonspecific Leydig cell toxicants.

VI. Future Perspectives and Research

Male reproductive physiology has made great strides in recent years in improving our understanding of the control of spermatogenesis and the complex cellular interactions that occur within the testis. In relation to other organs, our knowledge of basic testicular biochemistry is relatively scant, although steadily increasing. It certainly has reached the stage where some closer interaction between those involved in the toxicity of chemicals to the testis and the testicular physiologists should occur with benefits to both fields. The toxicologists can utilize improved physiological understanding to probe the mode of action of compounds, and by the same token physiologists can take note of the specific effects being produced by chemicals for their possible use as probes to understand normal function.

The interest in pyruvate and lactate production by Sertoli cells and their utilization by germ cells has been taken up by a number of laboratories investigating the toxic effects of compounds, where a number of commonalities are showing through. This interaction appears paradoxical at first sight, in that the Sertoli cells with a closer position to the blood supply and oxygen (see Fig. 3) extensively metabolize glucose anaerobically to produce lactate. The germ cells, some of which are four cell layers removed from the blood supply, cannot metabolize glucose anaerobically sufficiently to maintain their ATP levels, yet they can convert lactate or pyruvate aerobically to ATP. It seems likely that these and other cellular interactions will be an area of increased interest for both toxicologists and physiologists.

The relationship between discrete testicular injury and resulting reproductive function and outcome is also an area worthy of increased research. Our first impressions are that specific damage to different testicular cell types will have large differences in outcome, but our data base is small and our end points are crude. Such studies could be valuable in our interpretation and understanding of the relevance of "minor" testicular changes in standard reproductive toxicity studies. It would seem likely that a testicular biochemical lesion may affect, for instance, sperm function (e.g., motility, fertilizing capacity) where major histopathological damage to the testis or sperm abnormalities are not apparent (for example, see Holloway *et al.,* 1986). More sensitive indices of function are required in addition to rodent mating studies to increase our understanding of the effects of chemicals on male gonad function (see Chapter 8).

These comments are not meant to portray a gloomy outlook, but present to the reader/student an interesting area of increasing concern worthy of further detailed research.

References

Barney, G. H., Orgebin-Crist, M. C., and Macapinlac, M. P. (1969). Genesis of esophageal parakeratosis and histologic changes in the testes of the zinc-deficient rat and their reversal by zinc repletion. *J. Nutr.* **95,** 526–534.

Bellvé, A. R., and Feig, L. A. (1984). Cell proliferation in the mammalian testis: biology of the seminiferous growth factor. *Rec. Prog. Mammal. Res.* **40,** 531–567.

Bergh, A. (1982). Local differences in Leydig cell morphology in the adult rat testis: evidence for a local control of Leydig cells by adjacent seminiferous tubules. *Int. J. Androl.* **5,** 325–330.

Bergh, A. (1985). Development of stage-specific paracrine regulation of Leydig cells by the seminferous tubules. *Int. J. Androl.* **8,** 80–85.

Blackburn, D. M., Lloyd, S. C., Gray, A. J., and Foster, P. M. D. (1985). Testicular toxicity of 1,3-dinitrobenzene. *Toxicologist* **5,** 121.

Boyer, T. D., and Kenney, W. C. (1985). Acidic glutathione-*S*-transferases of rat testis. *Biochem. J.* **230,** 125–132.

Chapin, R. E., Morgan, K. T., and Bus, J. S. (1983). The morphogenesis of testicular degeneration induced in rats orally administered 2,5-hexanedione. *Exp. Mol. Pathol.* **38,** 149–169.

Chapin, R. E., Dutton, S. L., Ross, M. D., Sumrell, B. M., and Lamb, J. C. (1984a). The effects of ethylene glycol monomethyl ether on testicular histology in F-344 rats. *J. Androl.* **5,** 369–380.

Chapin, R. E., White, R. D., Morgan, K. T., and Bus, J. S. (1984b). Studies of the lesions in the testis and epididymis of F-344 rats by inhaled methyl chloride. *Toxicol. Appl. Pharmacol.* **76,** 328–343.

Chapin, R. E., Dutton, S. L., Ross, M. D., and Lamb, J. C. (1985). Effects of ethylene glycol monomethyl ether on mating and epididymal sperm parameters in F-344 rats. *Fundam. Appl. Toxicol.* **5,** 182–189.

Chapin, R. E., Gray, T. J. B., Dutton, S. L., and Lamb, J. C. (1986). The effects of 2,5-hexanedione, acrylamide and mono-2-(ethylhexyl)phthalate on Sertoli cell-enriched cultures. *Toxicologist* **6,** 289.

Chubb, C., and Ewing, L. L. (1979). Steroid secretion by *in vitro* perfused testes: Testosterone biosynthetic pathways. *Am. J. Physiol.* **237,** E247–254.

Clark, R. V. (1976). Three-dimensional organisation of testicular interstitial tissue and lymphatic space in the rat. *Anat. Rec.* **184,** 203–218.

Connell, C. J., and Connell, G. M. (1977). The interstitial tissue of the testis. *In* "The Testis," (A. D. Johnson and E. R. Groves, eds.), Vol. IV, pp. 333–369. Academic Press, New York.

Creasy, D. M., and Foster, P. M. D. (1984). The morphological development of glycol ether-induced testicular atrophy in the rat. *Exp. Mol. Pathol.* **40,** 169–176.

Creasy, D. M., Foster, J. R., and Foster, P. M. D. (1983). The morphological development of *di-n*-pentyl phthalate-induced testicular atrophy in the rat. *J. Pathol.* **139,** 309–321.

De Kretser, D. M., and Kerr, J. B. (1983). The effect of testicular damage on Sertoli and Leydig cell function. *In* "Monographs in Endocrinology: The Pituitary and Testis" (D. M. de Kretser, H. G. Burger, and B. Hudson, eds.), pp. 133–154. Springer-Verlag, Berlin.

Dym, M., and Fawcett, D. W. (1970). The blood-testis barrier in the rat and the physiological compartmentation of the seminiferous epithelium. *Biol. Reprod.* **3,** 308–326.

Eik-Nes, K. B. (1975). Biosynthesis and secretion of testicular steroids. *In* "Handbook of Physiology" (D. W. Hamilton and R. O. Greep, eds.), Vol. 5, pp. 95–115. Williams & Wilkins, Baltimore, Maryland.

Ewing, L. L. (1983). Leydig cell. *In* "Infertility in the Male" (L. I. Lipshultz and S. S. Howards, eds.), pp. 43–69. Churchill, New York.

Fawcett, D. W. (1975). Ultrastructure and function of the Sertoli cell. *In* "Handbook of Physiology" (D. W. Hamilton and R. O. Greep, eds.), Vol. 5, pp. 21–55. Williams & Wilkins, Baltimore, Maryland.

Fawcett, D. W., Neaves, W. B., and Flores, M. N. (1973). Comparative observations on the intertubular lymphatics and the organization of the interstitial tissue of the mammalian testis. *Biol. Reprod.* **9,** 500–532.

Foster, P. M. D., Creasy, D. M., Foster, J. R., Thomas, L. V., Cook, M. W., and Gangolli, S. D. (1983). Testicular toxicity of ethylene glycol monomethyl and monoethyl ethers in the rat. *Toxicol. Appl. Pharmacol.* **69,** 385–399.

Foster, P. M. D., Foster, J. R., Cook, M. W., Thomas, L. V., and Gangolli, S. D. (1982). Changes in ultrastructure and localization of zinc in rat testis following the administration of *di-n*-pentyl phthalate. *Toxicol. Appl. Pharmacol.* **63,** 120–132.

Foster, P. M. D., Sheard, C. M., and Lloyd, S. C. (1986). 1,3-Dinitrobenzene: a Sertoli cell toxicant? *Excerpta Med. Int. Congr. Ser.* **716,** 281–288.

Foster, P. M. D., Lloyd, S. C., and Prout, M. S. (1987). Toxicity and metabolism of 1,3-dinitrobenzene in rat testicular cell cultures. *Toxicol. in Vitro* **1,** 31–37.

Fritz, I. B., Rommerts, F. F. G., Louis, B. G., and Dorrington, J. H. (1976). Regulation by FSH and cAMP of the formation of ABP in Sertoli cell-enriched cultures. *J. Reprod. Fertil.* **46,** 17–24.

Galdieri, M., Zana, B. M., Monaco, L., Ziparo, E., and Stefanini, M. (1983). Changes of Sertoli cell glycoproteins induced by removal of the associated germ cells. *Exp. Cell. Res.* **145,** 191–198.

Galdieri, M., Monaco, L., and Stefanini, M. (1984). Secretion of androgen binding protein by Sertoli cells is influenced by contact with germ cells. *J. Androl.* **5,** 409–415.

Goldberg, E. (1977). Isoenzymes in testes and spermatozoa. *In* "Isoenzymes: Current Topics in Biological and Medical Research" (M. C. Rattazzi *et al.*, eds.), pp. 79–124. Liss, New York.

Gray, T. J. B., and Beamand, J. A. (1984). Effect of some phthalate esters and other testicular toxins on primary cultures of testicular cells. *Food Chem. Toxicol.* **22,** 123–131.

Gray, T. J. B., and Gangolli, S. D. (1986). Aspects of the testicular toxicity of phthalate esters. *Environ. Health Perspec.* **65,** 229–235.

Grootegoed, J. A., Jansen, R., and van der Molen, H. J. (1984). The role of glucose, pyruvate and lactate in ATP production by rat spermatocytes and spermatids. *Biochem. Biophys. Acta* **767,** 248–256.

Grootegoed, J. A., Jansen, R., and van der Molen, H. J. (1985). Intercellular pathway of leucine catabolism in rat spermatocyte epithelium. *Biochem. J.* **226,** 889–892.

Guthenberg, C., Astrand, I. M., Alin, P., and Mannervik, B. (1983). Gluthathione transferases in rat testis *Acta Chem. Scand.* **37,** 261–262.

Hodgen, G. D., and Sherins, R. J. (1983). Enzymes as markers of testicular growth and development in the rat. *Endocrinology* **93,** 885–989.

Holloway, A. J., Foster, P. M. D., and Moore, H. D. M. (1986). Relationship between testicular function and fertility probed with EGME. *In* "Proceedings of IVth European Testis Workshop on Molecular and Cellular Endocrinology of the Testis University of Rome Capri, (M. Stefanini, ed.), p. 165.

Jackson, H., Fox, B. W., and Craig, A. W. (1961). Antifertility substances and their assessment in the male rodent. *J. Reprod. Fertil.* **2,** 447–465.

Jackson, H., Partington, M., and Fox, B. W. (1962). Effects of busulphan (myleran) on the spermatogenic population of the rat testis. *Nature (London)* **194,** 1184–1185.

Jutte, N. H. P. M., Grootegoed, J. A., Rommerts, F. F. G., and van der Molen, H. J. (1981). Exogenous lactate is essential for metabolic activities in isolated rat spermatocytes and spermatids. *J. Reprod. Fertil.* **62,** 399–405.

Jutte, N. H. P. M., Grootegoed, J. A., Rommerts, F. F. G., and van der Molen, H. J. (1983). FSH stimulation of the production of pyruvate and lactate by rat Sertoli cells may be involved in the hormonal regulation of spermatogenesis. *J. Reprod. Fertil.* **68,** 219–226.

Lacroix, M., Smith, F. E., and Fritz, I. B. (1977). Secretion of plasminogen activator by Sertoli cell-enriched cultures. *Mol. Cell. Endocrinol.* **9,** 227–236.

Leblond, C. P., and Clermont, Y. (1952). Definition of the stages of the cycle of the seminiferous epithelium of the rat. *Ann. N.Y. Acad. Sci.* **55,** 548–573.

Lei, K. Y., Abbasi, A., and Prasad, A. S. (1976). Function of pituitary-gonadal axis in zinc-deficient rats. *Am. J. Physiol.* **230,** 1730–1732.

Ludvigson, M. A., Waites, G. M. H., and Hamilton, D. W. (1982). Immunocytochemical evidence for the specific localization of aldose reductase in rat Sertoli cells. *Biol. Reprod.* **26,** 311–317.

Maines, M. D. (1984). Characterisation of heme oxygenase activity in Leydig and Sertoli cells of the rat testes—differential distribution of activity and response to cadmium. *Biochem. Pharmacol.* **33,** 1493–1502.

Meistrich, M. L., Trostle, P. K., Fraport, M., and Erickson, R. P. (1977). Biosynthesis and localisation of lactate dehydrogenase -X in pachytene spermatocytes and spermatids of mouse testes. *Dev. Biol.* **60,** 428–441.

Miller, G. E., Welsh, M. J., and Brabec, M. J. (1985). Alteration of lactate metabolism in Sertoli cells and spermatocytes by 1,2-dibromo-3-chloropropane. *Toxicologist* **5,** 120.

Mita, M., and Hall, P. F. (1982). Metabolism of round spermatids from rats: Lactate as the preferred substrate. *Biol. Reprod.* **26,** 445–455.

Parvinen, M. (1982). Regulation of the seminiferous epithelium. *Endocrine Rev.* **3,** 404–417.

Rommerts, F. F. G., Grootenhuis, A. J., Moogerbrugger, J. W., and van der Molen, H. J. (1985). Ethane dimethane sulphonate (EDS) specifically inhibits LH stimulated steroidogenesis in Leydig cells from mature rats but not immature rats. *Mol. Cell. Endocrinol.* **42,** 105–111.

Russell, L. D. (1977). Movement of spermatocytes from the basal to the adluminal compartments of the rat testis. *Am. J. Anat.* **148,** 313–328.

Russell, L. D. (1978). The blood-testis barrier and its formation relative to spermatocyte maturation in the adult rat: a lanthanum tracer study. *Anat. Rec.* **190,** 99–112.

Russell, L. D. (1980). Sertoli–germ cell interrelation: a review. *Gamete Res.* **3,** 99–112.

Russell, L. D., Malone, J. P., and Karpas, S. L. (1981). Morphological pattern elicited by agents affecting spermatogenesis by disruption of its hormonal stimulation. *Tissue Cell* **13,** 369–380.

Setchell, B. P. (1969). Do Sertoli cells secrete fluid into the seminiferous tubules? *J. Reprod. Fertil.* **19,** 391–392.

Setchell, B. P., and Waites, G. M. H. (1975). The blood-testis barrier. *In* "Handbook of Physiology," (D. W. Hamilton and R. O. Greep, eds.), Vol. 5, pp. 143–172. Williams &

Wilkins, Baltimore, Maryland.
Sikka, S. C., Swerdloff, R. S., and Rajfer, J. (1985). *In vitro* inhibition of testosterone biosynthesis by ketoconazole. *Endocrinology* **116,** 1920–1925.
Skinner, M. K., and Fritz, I. B. (1985). Androgen stimulation of Sertoli cell function is enhanced by peritubular cells. *Mol. Cell Endocrinol.* **40,** 115–122.
Skinner, M. K., and Griswold, M. D. (1980). Sertoli cells synthesize and secrete a transferrin-like protein. *J. Biol. Chem.* **255,** 9523–9525.
Steinberger, E. (1971). Hormonal control of mammalian spermatogenesis. *Physiol. Res.* **51,** 1–22.
Thompson, J. N., Howell, J. M., and Pitt, G. A. J. (1964). Vitamin A and reproduction in rats. *Proc. R. Soc. London B* **159,** 510–535.
Van Thiel, D. M., Gavaler, J. S., Cobb, C. F., Santucci, L., and Graham, T. O. (1983). Ethanol, a Leydig cell toxin: Evidence obtained *in vivo* and *in vitro. Pharmacol. Biochem. Behav.* **18,** 317–323.
Williams, J., and Foster, P. M. D. (1987). The effect of 1,3-dinitrobenzene and mono-2-(ethylhexyl)phthalate on lactate and pyruvate production by rat Sertoli cell cultures. *Toxicologist* **7,** 142.
Zirkin, B. R., Ewing, L. L., Kroman, N., and Cochran, R. C. (1980). Testosterone secretion by rat, rabbit, guinea pig, dog and hamster testes perfused *in vitro:* Correlation with Leydig cell ultrastructure. *Endocrinology* **107,** 1867–1874.

Chapter 3

Duct System and Accessory Glands of the Male Reproductive Tract

E. M. EDDY

Gamete Biology Section
Laboratory of Reproductive and Developmental Toxicology
National Institute of Environmental Health Sciences
National Institutes of Health
Research Triangle Park, North Carolina 27709

I. Introduction

The duct system of the male reproductive tract connects the testis to the urethra via the efferent ducts, epididymis, ductus deferens, and ejaculatory duct. Associated with this paired duct system are the accessory sex glands, consisting of seminal vesicles, prostate, and bulbourethral glands in humans, and the additional coagulating glands and preputial glands in rodents. The duct system and most accessory sex organs in humans and laboratory rodents (rats and mice) are quite similar in structure and appear to have many common functional properties.

The male reproductive tract and accessory sex organs have critical roles in the reproductive process. Spermatozoa that leave the testis are incapable of fertilizing ova until they interact with the initial segments of the male reproductive tract. In addition, secretory products of the epithelia lining the tract and the accessory sex organs are required to carry spermatozoa into the female reproductive tract and provide the milieu in which spermatozoa can survive and function. There have been only a few reports of environmental agents that have detrimental effects on the structure or function of the ducts of the male reproductive tract or accessory sex organs. However, it is not clear if this is because such effects seldom occur, or because they have not been detected. Some toxic effects on the tract or organs might not alter the parameters usually measured, such as sperm numbers or motility, but might have effects on sperm or seminal plasma components that are not usually measured and that have essential roles in the reproductive process.

This chapter has two purposes. One is to describe the normal structure and function of the male reproductive tract and accessory sex organs to provide a basis for identifying changes in response to toxic substances. The other is to describe biochemical markers of the male tract and organs that may serve as useful indicators of toxic effects. Changes in the amount of specific molecules in the semen may help to identify the target organs and to define the mechanisms of action of particular compounds. In addition, a better understanding of the role of particular molecules in reproductive processes may come from such studies.

II. Development of Reproductive Tract and Accessory Organs

Exposure of human and rodent embryos to toxic agents during critical periods of development can result in male genital-tract abnormalities. For example, *in utero* exposure to the potent synthetic estrogen diethylstilbesterol can elicit teratogenic changes in the fetus and result in carcinogenesis later during adult life (McLachlan, 1981; Newbold and McLachlan, 1985). The changes seen in experimentally treated mice include failure of the testes to descend into the scrotum, lack of growth and differentiation of the seminal vesicles, formation of epididymal cysts, and occasional appearance of adenocarcinoma of the rete testis. The detection of such changes requires a knowledge of the development of the reproductive tract and accessory organs.

Most of the ducts and accessory glands of the male reproductive tract are derived during embryogenesis from the mesonephros and the mesonephric (Wolffian) duct (Fig. 1a). The mesonephric excretory system is functional in amphibia and other lower vertebrates but is replaced by the metanephric kidney and ducts in mammals and other higher vertebrates. However, during mammalian embryogenesis a mesonephric system forms and then regresses. After the seminiferous cords develop and join at the rete testis, they connect with nearby mesonephric tubules, each of which formerly connected a vascular glomerulus to the mesonephric duct. These mesonephric tubules become the efferent ducts. Most other mesonephric tubules dorsal and ventral to the gonad degenerate, although remnants may remain as part of the paradidymis and appendix of the epididymis. The cranial part of the mesonephric duct degenerates, but the middle part adjacent to the gonad becomes the epididymis (Fig. 1b). This region of the mesonephric duct grows rapidly and folds upon itself during organogenesis, forming the coils of the epididymis (Narbaitz, 1974).

The lower part of the mesonephric duct becomes the ductus deferens and the ejaculatory duct. A lateral evagination occurs near the caudal end of this duct, forming the primordium of the seminal vesicle and defining the junction of ductus deferens and ejaculatory duct (Figs. 1b and 1c). The evagination branches into three to four connected ducts to form the glandular portion of the seminal vesicle.

The mesonephric duct empties into the urogenital sinus. This sinus is separated into dorsal and ventral compartments by a septum growing down from the cephalic end. The ventral compartment gives rise to the bladder and urethra, the dorsal part to the rectum (Fig. 1c). The prostate forms in the posterior wall of the ventral urogenital sinus, in the region where the mesonephric ducts enter (Fig. 1d). Solid cords of cells grow from the wall of the urogenital sinus, branch repeatedly, and later acquire a lumen to become the glandular tissue of the prostate. The paramesonephric (Müllerian) ducts, which give rise to most of the reproductive tract in the female, form only the prostatic utricle in the male, a small structure that lies between the paired openings of the ejaculatory ducts into the prostatic portion of the urethra (Fig. 1b).

In humans, cords of cells arise dorsal, lateral, and anterior to the urethra. It has been suggested that two of the prostatic cords are from a part of the urogenital sinus derived from the paramesonephric duct. These cords may give rise to glands in the central part of the prostate, the region where benign prostatic hypertrophy usually occurs. In rodents, three different groups of cords appear. A dorsal group appears first as two pairs of cords growing from the dorsal wall of the urogenital sinus. These give rise to the coagulating glands. Dorsal and ventral groups of cords appear later to give rise to the remainder of the prostate (Narbaitz, 1974).

In addition to these major accessory glands of the male reproductive tract, there are two other paired glands associated with the distal part of the male reproductive tract. The bulbourethral glands develop as endodermal outgrowths from the membranous part of the urethra in both humans and rodents. However, the preputial glands are present only in rodents. They are modified sebaceous glands and ectodermal in origin.

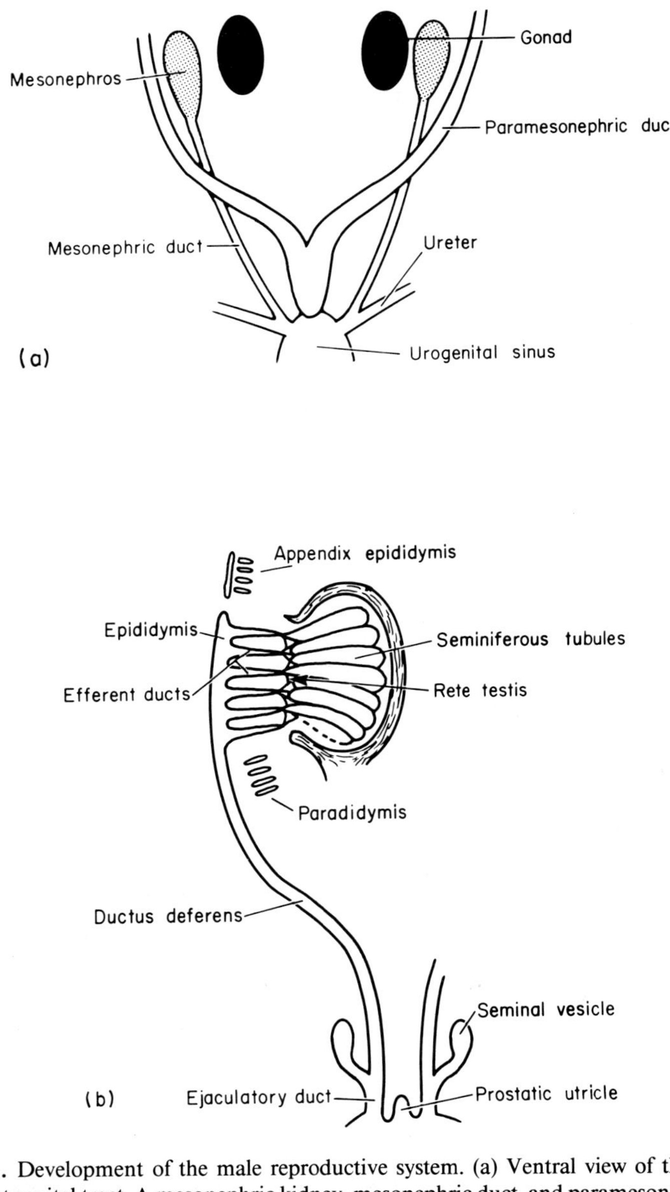

Fig. 1. Development of the male reproductive system. (a) Ventral view of the sexually indifferent genital tract. A mesonephric kidney, mesonephric duct, and paramesonephric duct are present on each side of the body before the male or female genital tracts develop. The paramesonephric duct regresses in the male, while the mesonephros and mesonephric duct contribute to the formation of the male reproductive tract and accessory glands. [Redrawn from Fitzgerald (1978).] (b) Development of the duct system. The efferent ducts are formed from tubules of the mesonephros and connect the rete testis to the mesonephric duct. The cranial part of the mesonephric duct becomes the epididymis, while the more caudal portion becomes the

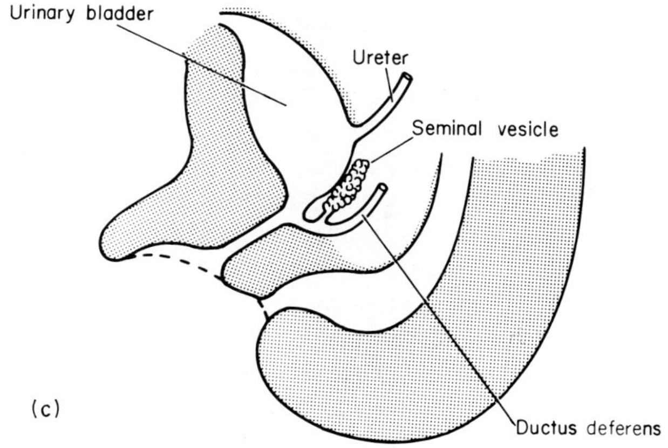

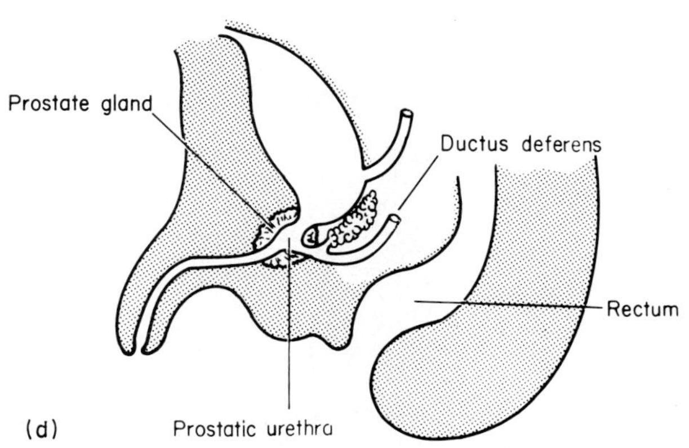

ductus deferens. The seiminal vesicles arise as lateral evaginations of the lower part of the mesonephric duct and define the junction between future ductus deferens and ejaculatory duct. [Redrawn from Langman (1963).] (c) Lateral view of developing accessory glands. The urogenital sinus separates into the rectum and the urinary bladder and urethra near where it is joined by the ejaculatory duct. The seminal vesicle lies dorsal to the urinary bladder and inferior to the ureter. [Redrawn from Langman (1963).] (d) Lateral view of developing accessory glands. The prostate gland develops in tissue of the posterior wall of the urogenital sinus and surrounds the prostatic urethra at the base of the urinary bladder. [Redrawn from Langman (1963).]

III. Structure and Function of Duct System

The duct system of the male reproductive tract is responsible for the transport and storage of spermatozoa after they leave the testis. Spermatozoa enter the duct system by passing from the rete testis (see Chapter 2) into the efferent ducts (see below) near the cranial pole of the testis (Figs. 1b and 2). Upon ejaculation, some spermatozoa from the distal part of the epididymis and those in the ductus (vas) deferens are discharged by smooth muscle contractions through the ejaculatory ducts.

The successful passage of spermatozoa through the excurrent duct system requires the coordinated action of nerves and smooth muscles of the duct

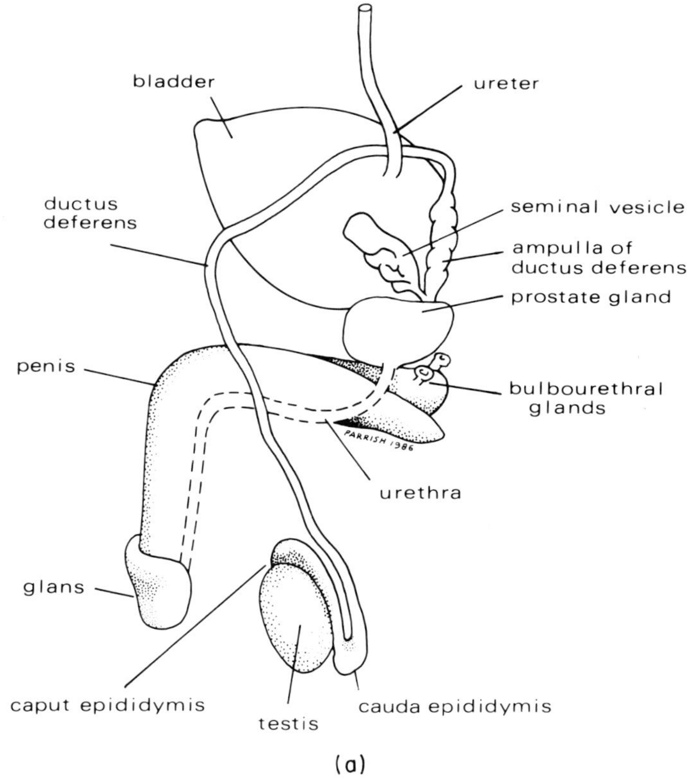

Fig. 2. The male urogenital system. (a) Lateral view of urogenital system of the human male. The duct system of the male reproductive tract connects the testis to the prostatic portion of the urethra and includes the epididymis, ductus deferens, and ejaculatory duct. Accessory glands of the male reproductive system include the seminal vesicles, prostate, and bulbourethral glands. The seminal vesicles join the duct system at the junction of the ampullar portion of the ductus deferens and the ejaculatory duct. The multiple ducts of the prostate enter the prostatic

and its blood supply and the appropriate function of physiological and biochemical processes carried out by the epithelial lining. Each of these activities and their regulatory mechanisms are potential targets for toxic agents that perturb male reproductive function.

A. Efferent Ducts

Twelve to fifteen efferent ducts (ductuli efferentes) pass from the rete testis to the head of the epididymis, where they join to form the single duct of the epididymis (Fig. 1b). The efferent ducts are arranged in cone-shaped spirals (coni vasculosi), with their bases toward the epididymis. In humans,

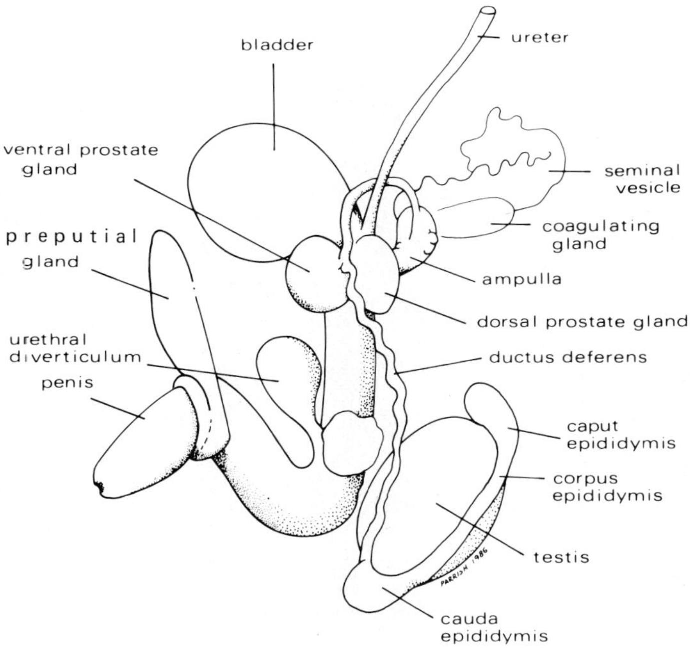

(b)

portion of the urethra, while the ducts of the paired bulbourethral glands enter the urethra near the base of the penis. (b) Lateral view of urogenital system of the male mouse. The duct system includes the epididymis, ductus deferens, and ejaculatory duct. Accessory glands of the male reproductive tract in rodents include the seminal vesicles, prostate, coagulating glands, and preputial glands. [Redrawn from Rugh (1968).]

the ducts are lined by alternate groups of low and tall columnar cells, which give the epithelium a scalloped appearance. In rodents, the epithelium is formed of low columnar cells (Fig. 3). The beat of the ciliated cells probably helps to move spermatozoa to the epididymis, while the more numerous nonciliated cells carry out secretory and absorptive functions (Hermo *et al.*, 1985). Both ciliated and nonciliated cells contain numerous granules in their apical cytoplasm. The epithelium is surrounded by a thin layer of circularly arranged, smooth muscle cells and loosely organized connective tissue.

B. Epididymis

The epididymis (ductus epididymis) lies closely opposed to the posterior–medial surface of the testis in humans and laboratory rodents

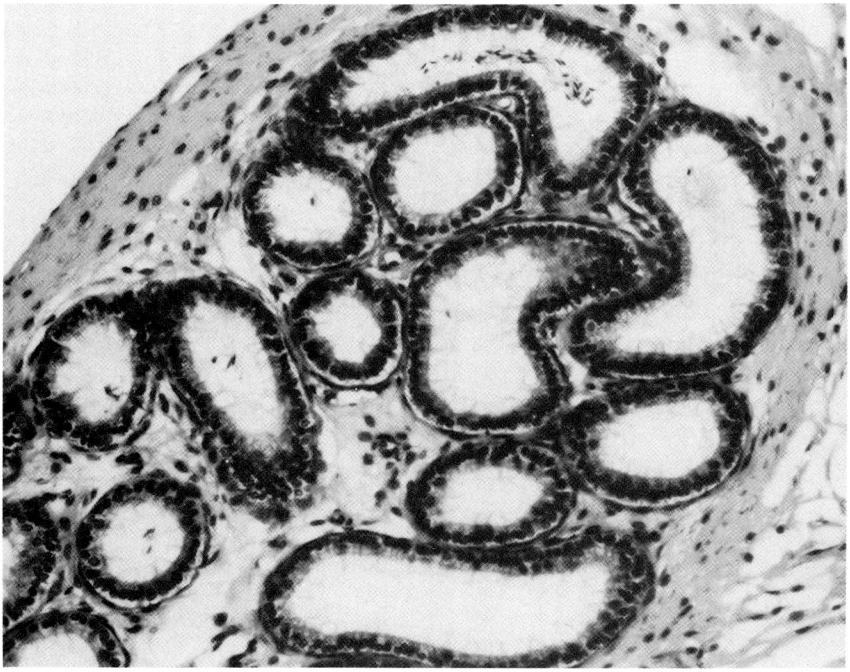

Fig. 3. Histology of efferent ducts of the adult mouse. The tissues shown in this and subsequent figures were Bouin's fixed, embedded in paraffin, sectioned, and then stained with hematoxylin and eosin. Twelve to 15 efferent ducts connect the rete testis to the caput epididymidis near the cranial pole of the testis. The low columnar epithelium is formed of nonciliated and ciliated cells. The nonciliated cells carry out secretory and absorptive functions, while the ciliated cells probably help to transport spermatozoa into the epididymis. The relative concentration of spermatozoa is low in the initial part of the duct system, and few sperm are seen in the ducts. × 200.

(Fig. 2). It consists of three main regions: the head (caput epididymidis), the body (corpus epididymidis), and the tail (cauda epididymidis). These regions may be further subdivided into a larger number of segments on the basis of morphology, vascularity, and histochemistry. The head of the epididymis is connected to the testis by the efferent ducts, and the tail is attached to the testis by connective tissue. The lateral surfaces of the head and tail are covered by serous membrane reflected from the surface of the testis, and the body is connected to the testis by a fold of the serous membrane. This serous membrane is the visceral layer of the sac of the tunica vaginalis, which was continuous with the peritoneal cavity in the fetus, but is separated in the adult.

The epididymis receives its blood supply and nerves via the spermatic cord. The rich blood supply of the epididymis is derived from two main sources. The testicular artery comes from the dorsal aorta and sends branches to the head of the epididymis. The external spermatic branch of the inferior epigastric artery anastomoses with the testicular artery. The lower part of the epididymis is supplied by a branch from the artery of the ductus deferens, which in turn comes from the superior vesical artery. Branches of the testicular and ductus deferens arteries anastomose along the epididymis (Clavert *et al.*, 1981). Epichlorohydrin and α-chlorohydrin have been reported to cause infertility by compromising the blood supply of the epididymis. They appear to damage vessels within the epididymis, in turn leading to damage of the lining epithelium (Waller *et al.*, 1985).

Veins from the epididymis join those emerging from the testis to form the pampiniform plexus. This network of veins around the spermatic cord serves as a countercurrent heat exchanger to cool arterial blood and to help maintain the temperature of the testis below that of the body. When these veins dilate to form varicocele, the stasis of blood flow can lead to sterility. This probably is because the elevated temperature damages specific stages of spermatogenesis (see Chapter 7). The autonomic nerves of the epididymis come from the lumbar plexus (sympathetic) and from the pelvic plexus (parasympathetic). Lymphatics draining the epididymis join those from the testis to form four to eight trunks that ascend through the spermatic cord to reach the lateral and preaortic groups of lumbar lymph nodes.

The epididymis is a single, highly coiled tubule. The intact epididymis is only a few centimeters long, but the total length of the tubule is up to 6 m in the human and 2 m in the rat. It is lined with a pseudostratified columnar epithelium containing two cell types. The *principal cells* are very tall in the head of the epididymis (Fig. 4) but gradually become shorter along the length of the tubule and are low columnar to cuboidal in the tail (Fig. 5). The diameter of the lumen increases along the epididymis, partly due to the decreasing height of the epithelium, but more due to the increasing size of the tubule (Fig. 5). The free surface of the principal cells is covered with long microvilli (stereocilia) and has numerous invaginations, while the cytoplasm contains a large supranuclear Golgi complex, abundant lysosomes, and numerous coated vesicles. These features are indicative of the active secretory and absorptive activity by the epithelium. These processes are critical for transport of sperm, modification of the composition of epididymal fluid, and changes in the sperm surface that occur during sperm maturation. Alteration of the function of the epididymal epithelium by toxic agents might affect any of these processes.

Also present within the epididymal epithelium are *basal cells,* small round or pyramid-shaped cells interdigitated between the bases of the principal cells. Some investigators have suggested that basal cells give rise to

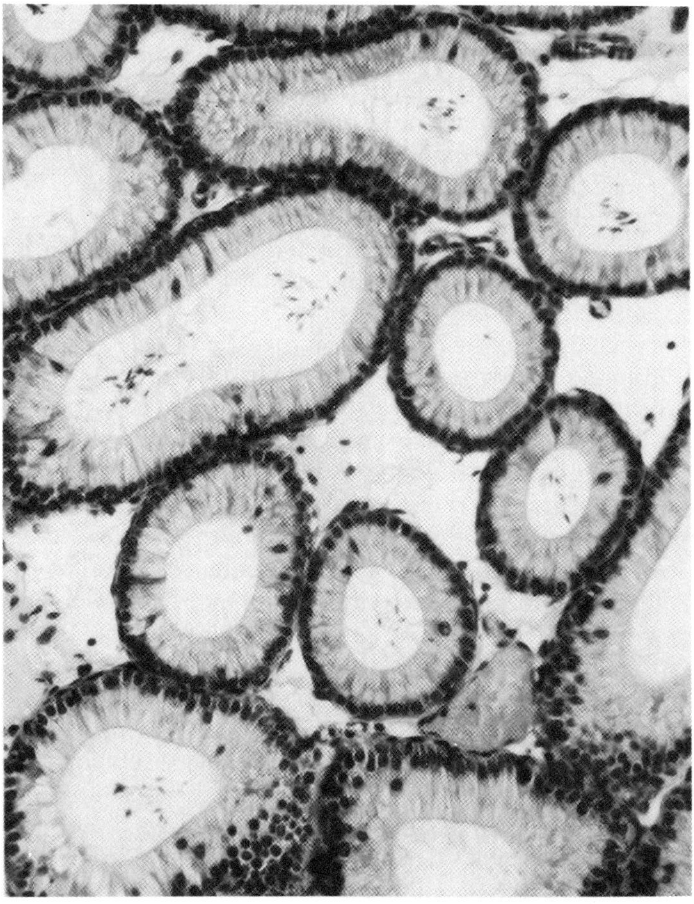

Fig. 4. Histology of caput epididymidis of the mouse. The single, highly coiled duct of the head of the epididymis is lined by tall columnar cells with long microvilli. × 200.

principal cells. In addition, lymphocytes are seen throughout the epithelium and were referred to previously as halo cells (Hamilton, 1975; Vendrely, 1981). Although halo cells are not frequent, they may have an important role in the immune system's tolerance of autoantigens present on the sperm surface.

A layer of smooth muscle fibers underlies the epithelium, with thin bundles of fibers encircling the proximal part and larger, oblique bundles of fibers surrounding the distal part of the epididymis. Sperm transport through

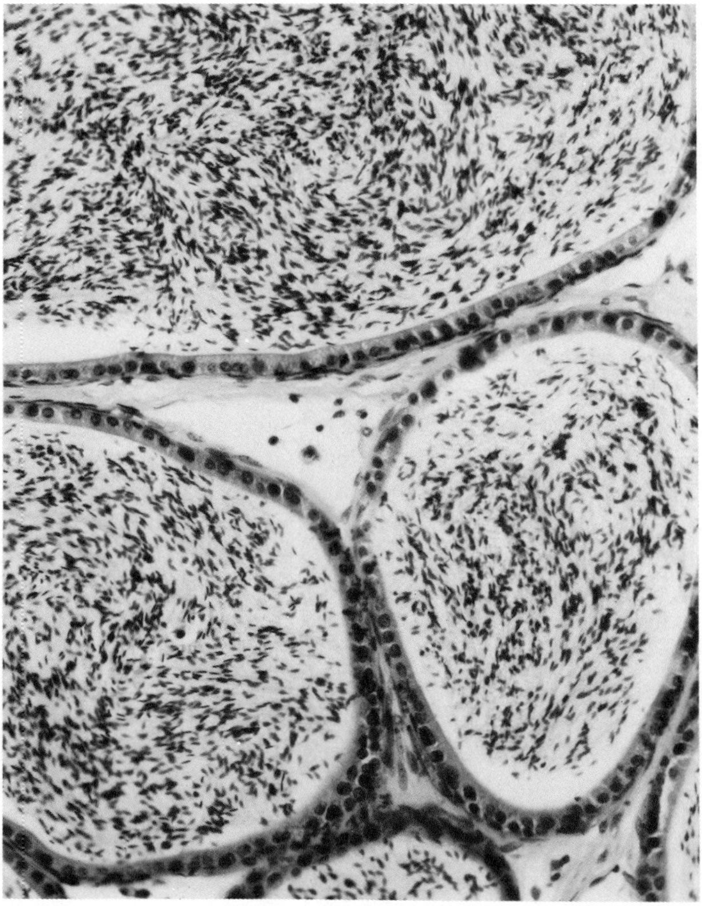

Fig. 5. Histology of cauda epididymidis of the mouse. The duct in the tail of the epididymis is lined by low columnar or cuboidal epithelial cells, and the diameter of the lumen is substantially greater than in the head of the epididymis. The concentration of spermatozoa is high in the distal portion of the duct system. × 200.

the epididymis continues for a few days following castration or efferent duct ligation, indicating that the movement is probably independent of hydrostatic pressure (Sujarit and Pholpramool, 1985). It appears that occasional weak spontaneous contractions in the proximal part of the epididymis move spermatozoa distally, while vigorous rhythmic contractions in the tail of the epididymis occur in response to stimulation by autonomic nerves to expel spermatozoa into the ductus deferens. Adrenergic nerves are numerous in the tail of the epididymis and terminate on smooth muscle cells, while those in the head of the epididymis are sparse and end on blood vessels. The prostaglandins are involved in regulation of contractility in the head of the epididymis and appear to modulate the stimulatory effects of norepinephrine, acetylcholine, and testosterone on contractions in the distal epididymis (Consentino *et al.,* 1984). This indicates that the smooth muscle component of sperm transport involves extrinsic nerve and endocrine signals and intrinsic processes, which work in concert to regulate contraction. Toxic agents that affect nerve, muscle, or endocrine function might in turn affect sperm transport.

1. Absorptive and Secretory Processes in the Epididymis

Besides being a reservoir for the storage of spermatozoa, the epididymis modulates the composition of the fluid in which spermatozoa are suspended by absorptive, ion transport, and secretory processes. Spermatozoa usually reach the tail of the epididymis 1 – 2 weeks after leaving the testis, but this varies among species and depends on frequency of ejaculation. The composition of epididymal fluid changes as it moves along the duct. Micropuncture sampling is the method of choice for collecting epididymal luminal contents free of tissue contaminants, but the small quantity of fluid obtained limits the types of analyses that can be carried out. However, this method is well developed and in experimental animals provides a potential approach both for assaying changes in epididymal fluid composition in response to toxic agents and for determining if such agents are present in epididymal fluid.

The epididymis absorbs more than 95% of the fluid coming from the testis. Sperm enter the epididymis as a dilute suspension, and only a few are seen in each tubule cross section in the head (Fig. 4). By the time they reach the tail of the epididymis, they are highly concentrated and fill the lumen (Fig. 5). During this process, there is a fall in epididymal fluid pH from 7.4 to 6.6 and a considerable change in the concentration of some ions. In the rat, rete testis fluid is similar to blood plasma in its high Na^+ and low K^+ concentrations, but fluid from the tail of the epididymis contains about a third as much Na^+ and about six times more K^+ than blood plasma. Although the combined ionic strength of cauda epididymal fluid is less than that of rete testis fluid or blood plasma, the osmolality of the fluid is maintained by organic substances such as glycerylphosphorylcholine and carnitine. The epididymal fluid also differs from rete testis fluid and blood plasma in concentration of other ions, enzymes, and organic components (Mann and Lutwak-Mann, 1981). It would be expected that inhibition of these processes by toxic agents would lead to changes in epididymal fluid composition, seminal volume, sperm count, and fertility.

The absorption of macromolecular and particulate material from the epididymal lumen has been demonstrated by introducing tracers such as α_2-macroglobulin, horseradish peroxidase, India ink, and colloidal iron (e.g., Djakiew *et al.,* 1985). These absorptive processes occur in the efferent ducts throughout the epididymis and in the ductus deferens. The tracers are first seen in pinocytotic vesicles at the base of microvilli, then in coated vesicles, and finally in multivesicular bodies and lysosomes deeper in the cytoplasm of principal cells. Under some experimental conditions, portions of degenerating spermatozoa are found in large multivesicular bodies within principal cells. However, only rarely has evidence of spermiophagy been reported in studies on normal epididymis (Hamilton, 1975).

The epididymal epithelium is also capable of actively transporting substances into the lumen. For example, the epididymis is unable to synthesize carnitine, but the concentration of carnitine in the lumen of the distal epididymis may be 2000 times that in the blood (Yeung *et al.,* 1980). This process is carried out by an androgen-dependent carnitine pump. This demonstrates the dependence of tissue-specific activities of the epididymal epithelium on endocrine function. The significance of carnitine in the epididymis has not been determined, but it has been proposed that it may serve both metabolic and osmotic roles. Additional studies are needed to determine if the low pH and the low sodium and high potassium concentrations in epididymal fluid are due to active transport processes, and, if so, how they are regulated.

The secretory activities of the epididymal epithelium alter the contents of the luminal fluid and modify both the biochemical composition and physiological characteristics of spermatozoa. Epididymal secretory products entering the luminal fluid include lipids, steroids, enzymes, and glycoproteins (Mann and Lutwak-Mann, 1981). Glycerylphosphorylcholine (GPC) is synthesized by the epithelium and is a major lipid constituent of epididymal fluid. It has been suggested that GPC has a protective or stabilizing effect on sperm and serves a role in maintaining osmotic pressure. The epididymal epithelium also appears to be a source of some of the steroids present in the luminal fluid. The epithelium contains hydroxysteroid dehydrogenases detectable by enzyme histochemistry and is able to produce testosterone and dihydrotestosterone *in vitro* from labeled precursors (reviewed in Hamilton, 1975). However, most of the androgens in the epididymal lumen enter with the rete testis fluid or are taken up from the circulation. Androgen-binding protein (ABP) produced in the testis (see Chapter 2) is present in the luminal fluid of the epididymis and is thought to be primarily responsible for the accumulation of dihydrotestosterone within the epididymis at 10–20 times the concentration of testosterone.

An active area of current research on the epididymis concerns identification of glycoproteins present in the luminal fluid and characterization of their synthesis and secretion (e.g., Flickinger *et al.*, 1984). Some of these glycoproteins have been shown to be androgen-dependent epididymal secretory products involved in maturation of spermatozoa. However, others may be shed from spermatozoa or epididymal epithelial cells, or produced by the testis, and not involved directly in epididymal function or maturation of spermatozoa. When these glycoproteins are better characterized, they may prove to be highly specific biochemical markers of epididymal function and useful indicators of target organ-specific toxic effects.

2. Epididymal Maturation of Spermatozoa

Spermatozoa gain the ability to fertilize as they transit the epididymis. During the time they are in the epididymis, spermatozoa undergo changes in morphology, composition, and function (reviewed by Orgebin-Crist *et al.*, 1975; Bedford, 1975; Hamilton, 1975). The morphological changes include loss of the cytoplasmic droplet and minor alterations in size and shape of the acrosome and midpiece. The changes in composition that have been described involve the appearance of new surface components and alterations in preexisting surface components. The functional changes include alterations in metabolism, gaining the ability to bind to the zona pellucida, and modifications in the pattern and effectiveness of flagellar activity. Although some of these changes clearly are essential for full sperm function, it has not been determined whether spermatozoa must undergo all of these changes to gain the ability to fertilize. Furthermore, the molecular processes underlying these phenomena have not been defined. However, methods are available for quantitating most of these functional processes, and they should be considered for their potential usefulness in monitoring toxic effects on the male tract and on sperm function.

In most mammals (rabbits appear to be an exception), spermatozoa are quiescent in the epididymis, and flagellar motion begins only after they are released from the epididymis. Spermatozoa removed from the efferent ducts or head of the epididymis and suspended in a physiological salt solution swim weakly with a vibratory movement that results in no forward progress. Some spermatozoa from the body of the epididymis are capable of moving more vigorously in a circular path, while the majority of spermatozoa isolated from the tail of the epididymis have a symmetrical flagellar beat, resulting in progressive forward movement.

There are multiple aspects to the process by which spermatozoa acquire the ability to carry out progressive forward movement. During epididymal maturation, there are increases in intracellular levels of cAMP and available cAMP-dependent protein kinase regulatory subunits. There is evidence that cAMP-dependent phosphorylation of a detergent-soluble protein (axokinin) initiates or maintains sperm motility (Tash *et al.,* 1984). There is also evidence of a heat-precipitable, enzyme-digestable "motility-inhibiting factor" in fluid from the tail of the epididymis of rat, bull, and rabbit (de Lamirande and Gagnon, 1984). This may be the same as immobilin, a glycoprotein of > 200,000 molecular weight present in epididymal fluid that mechanically immobilizes spermatozoa due to its high viscoelasticity (Usselman and Cone, 1983). Another glycoprotein component of epididymal fluid from several mammalian species, termed *forward motility factor,* is reported to bind to spermatozoa and to cause the induction of ability to carry out forward motility (Acott *et al.,* 1983). However, it also appears that in bulls and dogs the low intracellular pH of spermatozoa and low pH of epididymal fluid suppress motility of spermatozoa. Although it is not yet fully understood how these processes interact, it is likely that at least some of the regulatory mechanisms for forward motility vary among species (see Carr *et al.,* 1985).

High-speed video cameras linked to computer-based motion analysis systems are now commercially available that allow various parameters of sperm motion to be recorded and analyzed. This provides a quantitative and objective means for evaluating the direct or indirect effects of potentially toxic agents on sperm motility.

C. Ductus Deferens

The ductus (vas) deferens is a thick-walled, muscular tube. It begins at the distal end of the cauda epididymidis and passes from the scrotum into the abdominal cavity within the spermatic cord. At the beginning it is very tortuous but becomes less twisted as it ascends along the posterior body of the testis. The last part of the ductus deferens is dilated to form the spindle-shaped ampulla, which joins the duct of the seminal vesicle to become the ejaculatory duct. This pierces the body of the prostate to reach the prostatic part of the urethra (Fig. 2). Vasectomy usually involves ligation and separation of the scrotal portion of the ductus deferens and is a highly effective way to control male fertility. However, it can lead to granuloma formation and increased sperm antibody titers, and epididymitis is an occasional more serious complication.

The ductus deferens is lined with a pseudostratified, columnar epithelium (Fig. 6) with long microvilli (stereocilia) on the apical surface. The duct lumen is narrow, and the mucosa is thrown into longitudinal folds. The epithelial cells are secretory and are shorter than those in the epididymis. The nature and role of secretory processes in this organ have not been studied extensively, but they are probably like those in the epididymis. Some cells contain an extremely rich accumulation of mitochondria, suggesting that they may be active in ion transport. The epithelium is surrounded by spirals

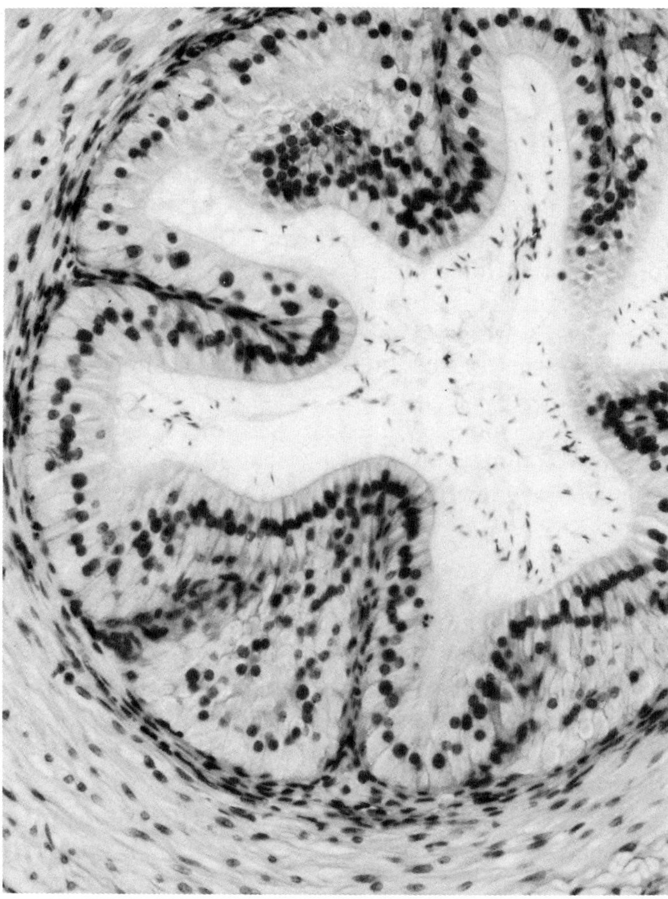

Fig. 6. Histology of ductus deferens of the mouse. The ductus deferens is lined with psuedostratified columnar epithelial cells with long microvilli. Longitudinal folds of the epithelium extend into the duct lumen, and the epithelium is surrounded by a thick layer of smooth muscle. × 200.

of smooth muscle, which make up most of the thickness of the wall of the epididymis. The mucosa of the ampulla is also arranged in longitudinal folds, which are branched and taller than those in the ductus deferens. The ampulla has a wider lumen and a thinner muscular coat than the ductus deferens, and glands extend deep into the muscular layer. The guinea pig and boar lack an ampulla. The ejaculatory duct is lined by simple or pseudostratified columnar epithelium, which become transitional epithelium near the opening into the urethra. It has a smaller lumen than the ductus deferens, and the lining is folded longitudinally.

IV. Structure and Function of Accessory Glands

Most of the volume of the ejaculate consists of secretory products of the accessory sex glands. The roles of few of these products are known, but they can serve as useful qualitative and quantitative indicators of reproductive-tract and accessory sex-gland function (Table I). Because several of these are produced by particular accessory glands, they have the potential to be used as markers for monitoring target organ-specific effects of toxic substances. *In vitro* fertilization in rodents can be carried out with spermatozoa from the cauda epididymidis in the absence of accessory-gland secretory products, indicating that these products are not required for processes directly involved in rodent sperm–egg fusion. However, fertility is reduced *in vivo* in the absence of accessory-gland products, indicating that they contribute to the overall processes of reproduction.

A. Seminal Vesicles

The seminal vesicles are paired, lobulated glands lying dorsal to the urinary bladder (Fig. 2). The arteries supplying the seminal vesicles are derived from the middle and inferior vesical and middle rectal arteries, all of which arise from the internal iliac artery. The veins and lymphatics accompany the arteries. The nerves are derived from the pelvic plexus. The seminal vesicles are elongated, hollow organs with small saclike evaginations in the walls. The walls also contain inner–circular and outer–longitudinal layers of smooth muscle and a rich accumulation of elastic fibers. The epithelium is thrown into tall, thin folds (Fig. 7), which branch and anastomose to form irregular side chambers that communicate with the central chamber. The epithelium is pseudo-stratified with tall columnar cells and basal cells similar to those seen in the epididymis. The epithelium has the appearance characteristic of active protein secretion, the cells containing abundant rough endoplasmic reticulum, vacuoles, and dense secretory granules (Brandes, 1974). The gen-

TABLE I

Secretory Products of Ducts and Accessory Organs of Male Reproductive Tract

	Humans	Rodents
Epididymis	Carnitine Glycerophosphorylcholine	Carnitine Glycerophosphorylcholine Motility-inhibiting factor Immobilin Forward motility factor AEG, SEP, etc.
Seminal vesicles	Citric acid Fructose Lactoferrin Proteinase inhibitors HUSI I and II	Citric acid Fructose Basic secretory proteins Proteinase inhibitors Coagulinogen
Prostate	Transferrin Prostatein Zinc Citric acid Acid phosphatase Seminal liquification enzymes Inositol Spermine	Transferrin Prostatein Dorsal proteins I and II Fructose
Coagulating glands	—	Vesiculase Fructose
Bulbourethral glands	"Lubricant"	Coagulating factor
Preputial glands	—	β-Glucuronidase Pheromones
?	Immunosuppressive factors Viruses	Immunosuppressive factors Viruses

eral organization and structural features of the seminal vesicles are quite similar in humans and laboratory rodents, but the glands are relatively larger in rodents than in primates.

The secretion of the seminal vesicles is a viscous substance that makes up 50–80% of the volume of the ejaculate in man (Zaneveld and Tauber, 1981). The secretion is rather alkaline and thought to aid in neutralizing the acid pH of vaginal fluid. It is rich in fructose, and levels of this constituent in the semen are used for evaluating seminal vesicle function in men. In mice and rats, fructose appears to be produced mainly by the coagulating glands and prostate. Citric acid is a major component in seminal vesicle secretions in humans and rodents, while inositol is found at high levels only in rodents. The iron-binding protein, lactoferrin, is produced by the seminal vesicles in the human and becomes one of the "sperm-coating antigens" in the ejaculate (see Mann and Lutwak-Mann, 1981). A group of small, basic secretory proteins is produced in rat and mouse, and the genes for two of these, seminal vesicle secretion proteins IV and V, have been cloned in the rat (Harris *et al.*, 1983). A major protein in seminal plasma and coating spermatozoa from normal men is immunologically related to rat seminal vesical protein IV (Abrescia *et al.,* 1985). Surgical removal of the seminal vesicles results in reduced fertility in rats, indicating that at least some of these secretory products are important in reproduction. It should be possible to determine the response of this gland to toxic agents by measuring the relative concen-

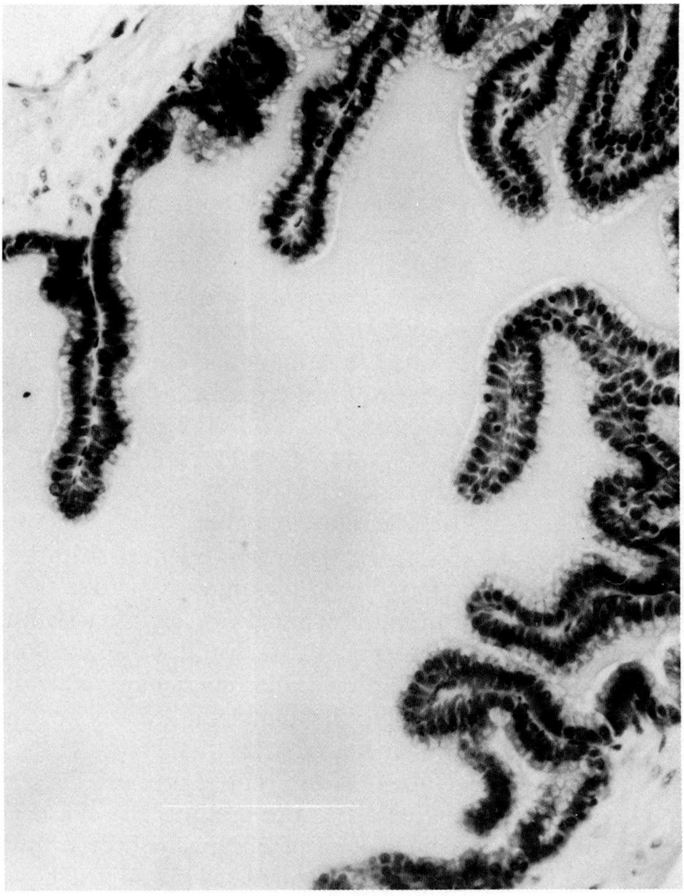

Fig. 7. Histology of seminal vesicle of the mouse. The seminal vesicles are lined by pseudostratified cells, and the epithelium is thrown into tall folds, which branch and anastomose. The lumen contains homogeneously staining secretory material. × 200.

trations of these different secretory products in the semen, using specific antisera and radioimmunoassays.

B. Prostate

The prostate lies immediately caudal to the urinary bladder, surrounding the first part of the urethra (Fig. 2). The arteries supplying the prostate are derived from the internal pudendal, inferior vesicle, and middle rectal arteries, all branches of the internal iliac artery. The veins form a plexus around the sides and base of the gland and enter the internal iliac veins. The nerves come to the prostate from the pelvic plexus. In the human the prostate is a compact organ, containing three groups of glands within a common fibromuscular capsule. The *periurethral mucosal glands* immediately surround the urethra, and their ducts empty directly into the prostatic urethra. This portion of the prostate contributes a minor part of the secretions but is the location of benign prostatic hyperplasia in the majority of men over the age of 60. Most of the "true" prostate is in the peripheral zone, containing *submucosal glands* and *prostatic glands.* This portion of the prostate contains 30–50 tubuloalveolar glands, which empty into the posterolateral urethral sinus by 15–30 ducts. Prostatic carcinoma in men usually occurs in this peripheral part of the prostate. The glands form a lobe on each side of the urethra and are surrounded by a stroma containing smooth muscle. The glands and ducts are lined by simple columnar and pseudo-stratified columnar epithelium (Fig. 8), with transitional epithelium being found in ducts near their opening into the urethra. The cells have abundant rough endoplasmic reticulum and numerous vacuoles and granules, indicative of their active secretory role. Prostatic concretions within the glandular alveoli are thought to be condensations of secretory material.

The *coagulating glands* of the rat and mouse are derived from the prostate and sometimes have been referred to as the anterior lobes of the prostate. They lie immediately adjacent to the seminal vesicles and within the same peritoneal sheath as these glands (Fig. 9). The gland is composed of tubular structures lined with cuboidal and columnar epithelium, which is deeply folded (Fig. 10). The epithelium appears to undergo secretory cycles, resulting in the cuboidal and columnar shaped cells.

The prostate in rodents consists of the ventral prostate, present as two pear-shaped structures lying ventral to the urethra, and the *dorsolateral prostate,* formed of a medial and two lateral lobes encircling the remainder of the urethra (Rugh, 1968). The prostate contributes 15–30% of the volume of the ejaculate. The prostatic secretion is a colorless fluid rich in proteolytic enzymes, zinc, citric acid, and acid phosphatase (see Waller *et al.,* 1985). In addition, uteroglobin has been identified immunologically in the epithelium

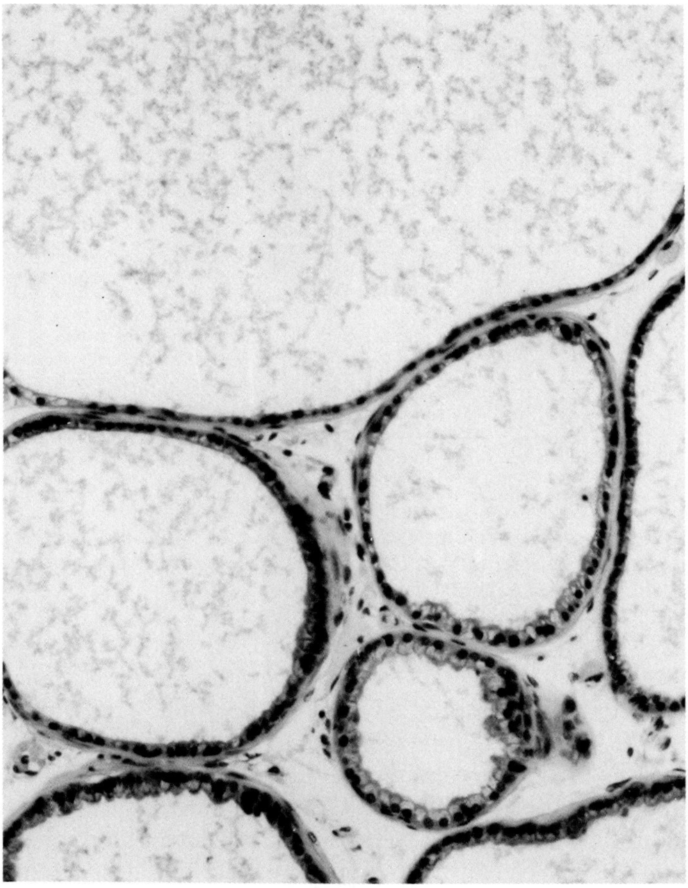

Fig. 8. Histology of prostate of the mouse. The glands of the prostate are lined by low cuboidal or columnar epithelial cells, with the height depending on secretory activity. The lumen contains a floculant precipitate of secretory material. × 200.

of the dorsolateral prostate in the rabbit (Muller, 1983). The prostate also contains transferrin, detectable by immunocytochemistry in the dorsal and lateral prostate but not in the coagulating gland. Transferrin is elevated in patients with prostatic tumors, and the Dunning rat prostate tumor produces transferrin at high levels (Wilson *et al.,* 1981). The coagulating glands in rodents produce an enzyme required for the formation of the copulatory plug (see below).

Prostatein, the major protein present in the rat ventral prostate (Wilson *et al.,* 1981), binds various nonpolar hormonal steroids and synthetic estradiol

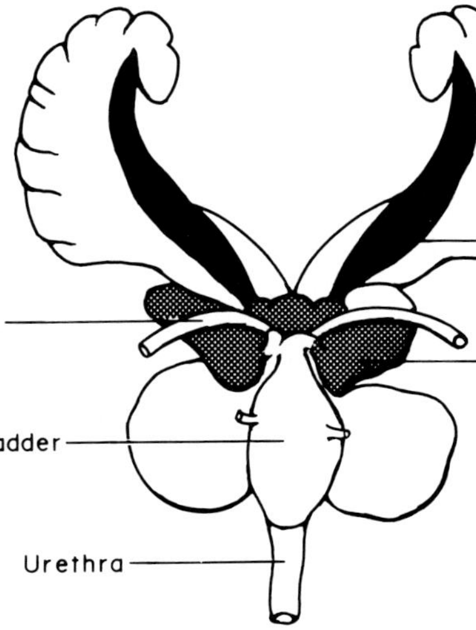

Fig. 9. Accessory glands of the male reproductive tract in the rat. Shown is a ventral view of the seminal vesicles and closely associated coagulating glands, the prostate, and portions of the urinary system. [Redrawn from Mann and Lutwak-Mann (1981).]

derivatives. It has also been called rat prostatic-binding protein, α-protein, and estramustine-binding protein by different investigators. It has a native molecular weight of approximately 42,000 and is composed of 20,000 and 22,000 molecular weight subunits. These subunits in turn are each composed of disulfide cross-linked components. Prostatein binds another small protein (~ 5200 molecular weight) rich in proline, and in combination they have a higher pregnenalone-binding affinity than the major protein does alone. Prostatein appears during juvenile life, and synthesis is androgen dependent. It is thought to serve as a carrier function for steroids, but it may also be a seminal protein necessary for copulatory-plug formation. A similar protein has been reported in the human prostate.

Two proteins have been purified from the rat dorsal prostate. Dorsal protein I is about 25% of the cytosolic protein and 6% of the protein in prostatic fluid. It is a 150,000 molecular weight dimer, composed of two 71,000 molecular weight monomers, and has a low carbohydrate content. Dorsal protein II is androgen dependent and increases after 20 days of age. It is around 300,000 molecular weight in the native form and about 80,000 following reduction of disulfide bonds. It is less abundant than dorsal protein I and has a high carbohydrate content (Wilson *et al.*, 1981).

C. Bulbourethral Glands

The bulbourethral (Cowper's) glands are small paired organs on each side of the urethral bulb. They are connected to the penile urethra by long ducts. They contain compound, tubuloalveolar glands lined by cells varying greatly in height and appearance. The cells appear to cycle from a resting state, when they are low columnar and contain granular cytoplasm, and to a secretory state, when they are tall columnar and filled with mucus. The glands are divided into small lobules by septa of connective tissue and smooth muscle.

The bulbourethral gland produces a small volume of a clear, viscous, mucuslike substance. In the human, it is released before ejaculation and may be a lubricant, while in rodents it probably contains a clotting agent involved in copulatory-plug formation (see below).

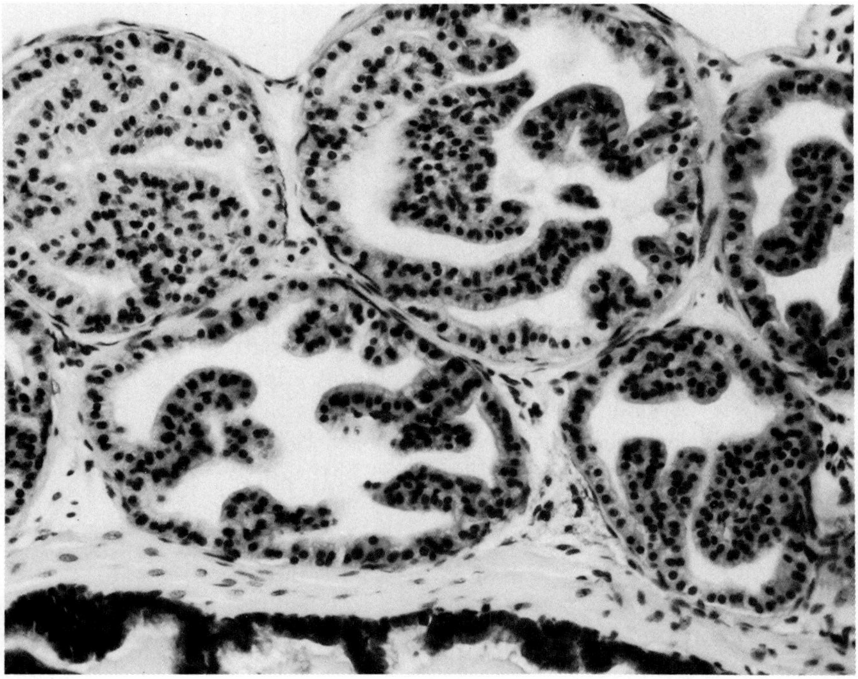

Fig. 10. Histology of coagulating gland of the mouse. The coagulating glands are formed of tubules lined with columnar epithelial cells. Folds of the epithelium extend into the lumen, and the height of the epithelium varies with the secretory activity in different regions of the gland. A portion of the adjacent seminal vesicle is present at the bottom. × 200.

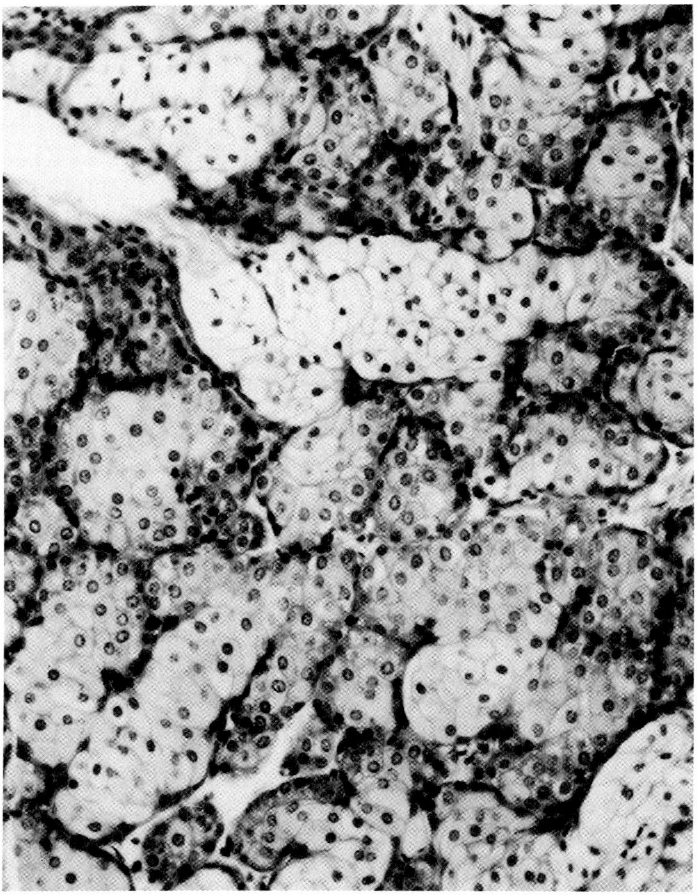

Fig. 11. Histology of preputial gland of the mouse. The lobules of the preputial gland contain closely packed sebaceous cells. The holocrine secretion of these cells occurs into ducts lined with stratified epithelium. × 200.

D. Preputial Glands

These are large paired glands lying between the skin and the abdominal wall in mice and rats. They are modified sebaceous glands (Fig. 11) and open onto the surface on either side of the urethral meatus in males and females. Their secretions are rich in β-glucuronidase and possess a pheromonal activity. In rats, the pheromone is a volatile, lipid-soluble material containing aliphatic alcohols.

E. Endocrine Regulation of Ducts and Accessory Sex Glands

Androgens are required for the development and function of the male reproductive tract and accessory sex organs. Castration causes dramatic decreases in accessory sex organ weight and in RNA and protein synthesis, and this is usually fully reversible by testosterone treatment. However, androgens have relatively little effect on steady-state mRNA levels or overall protein synthesis in the epididymis compared to the effects on accessory sex organs (Brooks, 1983a,b).

A major difference between epididymis and accessory sex organs is the amount of synthetic activity devoted to producing secretory proteins. It has been estimated that only 15% of the total protein synthesis in the epididymis produces secretory proteins. Also, many proteins are present in epididymal secretion with no single protein predominating. However, in rat prostate, up to 40% of the total protein synthesis may involve a single protein, and two secretory proteins probably account for 25% of protein synthesis in the seminal vesicles.

Epididymal development depends on the presence of androgens. The mesonephric duct does not form in castrated male rabbit fetuses, while grafting a testis or inserting testosterone crystals into female rabbit fetuses before sexual differentiation causes mesonephric-duct development (Narbaitz, 1974). Furthermore, after castration or hypophysectomy, the adult epididymis shows decreased tubular diameter and epithelial height, histochemically detectable changes in enzymatic activity, and dramatically decreased carnitine content. Castration also results in accelerated transport of spermatozoa through the epididymis (Sujarit and Pholpramool, 1985) and causes epididymal spermatozoa to lose their ability to become motile when placed in physiological media. However, rat sperm in the proximal cauda epididymidis lose their ability to fertilize before motility is affected, and testosterone treatment prevents the rapid decrease in fertility. This is probably not directly due to the presence of androgens, because addition of an androgen-free epididymal protein extract increases the ability of immature hamster spermatozoa to fertilize *in vivo* and *in vitro* (Gonzales-Echeverria *et al.,* 1984). Therefore, sperm maturation in the proximal caput epididymidis appears to depend on synthesis of androgen-dependent secretory proteins.

It also has been found that after castration sperm held in the proximal caput epididymidis fail to gain the ability to bind to the zona pellucida. Sperm–zona binding is a surface phenomenon, and biochemical and immunological studies have demonstrated that substantial changes occur in the sperm surface during epididymal maturation (reviewed by Eddy *et al.,* 1985). In several species, including humans and rodents, glycoproteins are added to the sperm surface during epididymal maturation. In the rat, different investigators have reported that during transit through the epididymis spermatozoa acquire an "acidic epididymal glycoprotein" (AEG), "specific epididymal glycoproteins" (SEP), "proteins B, C, D, and E," a sialoprotein, and a 32,000 molecular weight protein. It has been suggested that the glycoproteins in many of these studies are probably the same molecules (Brooks, 1983a,b). Antibodies against androgen-dependent epididymal glycoproteins block fertilization in the rat as tested by artificial insemination with antibody treated spermatozoa (Cuasincu *et al.,* 1984). In some cases, production of these glycoproteins has been shown to be androgen dependent (Fournier-Delpech *et al.,* 1984). Based on such findings, it appears that androgen-dependent glycoproteins secreted by the epididymal epithelium are added to the sperm surface during maturation to produce zona-binding ability (Saling, 1982; Orgebin-Crist and Fournier-Delpech, 1982). However, a glycoprotein has not been identified that mediates sperm – zona binding, and the mechanisms involved in the production, sperm surface attachment, and activity of this molecule have not been determined.

Castration, hypophysectomy, or treatment with antiandrogens causes rapid and substantial atrophy of the seminal vesicles and the prostate. The rate of atrophy is often greater after hypophysectomy than after androgen deprivation, suggesting that androgens are not solely responsible for regulating function of the male accessory sex organs. The seminal vesicles of long-term castrated rats contain cuboidal epithelial cells that lack the extensive rough endoplasmic reticulum, large Golgi complexes, and abundant secretory granules present in intact rats. Testosterone treatment of castrated or hypophysectomized rats results in recovery of the seminal vesicle epithelium in 3 – 5 days. Castration also results in decreased epithelial height and secretory activity in the prostate, with involution becoming apparent at the electron-microscope level after 2 days. By 4 days after castration, the organ weight and protein content of the prostate decrease to less than one-fifth of their initial values. However, these structural and functional changes are rapidly reversed by androgen treatment even weeks after regression.

Because the development of the male reproductive system and the synthesis of many of the secretory products of the duct system and accessory glands are androgen dependent, toxic agents that alter androgen production are likely to have a profound effect on male reproductive function. However, some agents might influence hormone action at the level of the receptor or by altering the intracellular response in a tissue- or cell-type specific manner. Consequently, defining the mechanism of action of some agents may require monitoring the production of specific molecules, even though they are among the androgen-dependent secretory products.

V. Seminal Plasma

Semen is composed of spermatozoa and seminal plasma. Seminal plasma is one of the few body fluids that can be obtained noninvasively and can be used to monitor concentrations of toxic agents in potentially exposed human males (Hales *et al.*, 1986). Of course, seminal plasma can also be used to determine the effects of these agents on specific regions of the tract and on different accessory organs in experimental animals, using the specific markers indicated above. In addition, there are other components in the seminal plasma whose origins have not been determined but that may nevertheless be useful to toxicologists studying male reproductive function.

A. Enzyme Inhibitors in Seminal Plasma

Inhibitors of trypsin-like enzymes have been identified in the seminal plasma of mammals. Two human seminal plasma trypsin inhibitors (HUSI I and II) are acid-stable basic proteins of approximately 6000 and 12,000 molecular weight (see Zaneveld and Chatterton, 1982). HUSI I inhibits trypsin and chymotrypsin, and HUSI II inhibits trypsin and acrosin. In addition, two acid-stable proteinase inhibitors of 4000 and 6400 molecular weight have been isolated from mouse reproductive tract tissues. The smaller inhibitor is produced in the epididymis and is a competitive inhibitor of trypsin, while the larger inhibitor comes from the seminal vesicles and is a noncompetitive inhibitor of trypsin. The seminal vesicle inhibitor binds to a 15,000 molecular weight acceptor molecule on the plasma membrane over the acrosome. An antiserum against the seminal vesicle inhibitor does not cross-react with the epididymal inhibitor (Poirier and Jackson, 1981; Aarons *et al.*, 1984). These inhibitors might play a role in the regulation of acrosin activity during fertilization or may protect the male and female reproductive tracts from the effects of proteases released from damaged spermatozoa or produced by the accessory sex glands.

B. Coagulation of Semen

Human semen undergoes coagulation immediately after ejaculation and then liquefaction in 10–20 min. Coagulation appears to involve interaction between a clotting enzyme and a coagulable protein from the seminal vesicles, neither of which has been isolated. Liquefaction is believed to be caused by two proteolytic factors from the prostate, including a collagenase-like enzyme and a chymotrypsin-like enzyme (seminin). Also involved are a plasminogen activator and a plasminogen pro-activator. The activators are approximately 69,000 and 74,000 molecular weight and appear to be similar to urokinase. Spermatozoa are trapped within the coagulum in the vagina and become motile when liquefaction occurs (Mann and Lutwak-Mann, 1981; Waller *et al.*, 1985).

The ejaculates of rats, mice, and guinea pigs consist of a first bolus of fluid containing sperm and a second bolus of accessory gland fluids that form a copulatory plug in the vagina, preventing loss of sperm and further copulation. This is believed to involve conversion of procoagulase, produced by the seminal vesicles, to coagulase by action of the enzyme vesiculase produced by the coagulating gland. This in turn acts on coagulinogen from the seminal vesicles to form the plug. The bulbourethral gland also produces a coagulating agent. Plug formation involves a transglutaminase activity, and polyamines in the semen are incorporated into cross bridges that result in the formation of a highly insoluble, high-molecular-weight protein complex. It has not been determined which, if any, of these components correspond to the secretory proteins discussed above.

A trypsin-like enzyme of approximately 29,000 molecular weight forms over 90% of the total protein in canine seminal plasma. This proteolytic enzyme is produced in the prostate and binds to the surface of the tail and postacrosomal regions of ejaculated spermatozoa (Isaacs and Coffey, 1984).

C. Immunosuppression by Seminal Plasma

Some sperm surface molecules are highly immunogenic, and antibodies to these components can be a cause of infertility. However, spermatozoa in the male or female reproductive tracts usually are not prone to attack by the immune system. Tolerance to most "self" components probably occurs during the perinatal period by elimination of self-reactive clones, but spermatogenesis begins long after this period. The presence of a blood–testis barrier within the seminiferous epithelium (see Chapter 2) and suppressor T-cells within the epididymal epithelium probably are important for allowing sperm autoantigens to exist within the male reproductive tract. There also appears to be a partial blood–epididymal barrier, but immunoglobulins from the blood plasma have been found in epididymal fluid in low concentrations (Weininger *et al.,* 1982). Vasectomy commonly leads to elevated titers of antisperm antibodies, apparently due to leakage of antigens across these barriers. However, the seminal plasma also contains immunosuppressive factors that cause a reduction of immune responses (James and Hargreave, 1984).

Factors in the seminal plasma can interfere with the *in vitro* function of T-cells, B-cells, natural killer (NK) cells, and macrophages. They can also impair the activity of antibody and complement molecules and decrease the ability of immune system cells to bind antigen, differentiate or proliferate, perform phagocytosis, or effect target-cell lysis. The factors have been demonstrated in several species, and the activities are not species specific. They are effective in quite small concentrations and appear to involve secretions of the epididymis, prostate, and seminal vesicles. Seminal plasma also has been reported to have immunosuppressive effects *in vivo.* Pooled extracts of epididymis, seminal vesicles, and prostate inhibited the primary and secondary responses in mice to foreign antigens and to epididymal spermatozoa. Immunosuppressive activity is found in semen samples from vasectomized males, indicating that at least some of the factors are not from testis, epididymis, or spermatozoa.

It appears that the immunosuppressive effects of seminal plasma are due to a variety of factors of different molecular weights, binding specificities, heat stability, and other different physical and chemical properties. Possible immunosuppressive factors that have been reported in seminal plasma include zinc–peptide complexes, transglutaminase, a 94,000 molecular weight Fc receptor-binding protein, a protein similar to pregnancy-associated protein A, polyamines, prostaglandins, lactoferrin, and β_2-microglobulin (for review, see James and Hargreave, 1984). Lactoferrin is produced by the seminal vesicles in humans, but the site of origin and specific activities of most other putative immunosuppressive molecules remain to be determined. Mechanisms that have been proposed for the immunosuppressive activity include masking or modification of sperm surface antigens to impair their recognition by immunocompetent cells, direct effects on immune-system cells themselves to reduce their ability to recognize and respond to antigens, and inhibition of binding of sperm-surface antigens to Fc receptors or phagocytic cells.

In addition to the potential roles of immunosuppressive factors in reproductive processes, they may have possible relevance in sexually transmitted diseases, local infections in the reproductive tract, and the presence of viruses associated with malignancy. It has been suggested that immunosuppressive factors in the seminal plasma may inhibit defense mechanisms to infective agents, such as those involved in transmission of gonorrhea, syphylis, and chlamydia and causing prostatitis and epididymitis. The viruses associated with cervical carcinoma and the acquired immune deficiency syndrome (AIDS) have been identified in seminal plasma (James and Hargreave, 1984; Ostrow *et al.,* 1986).

D. Drugs and Toxic Agents in Seminal Plasma

A variety of drugs and toxic agents has been identified in semen. This is of interest for two reasons. One is that it indicates that other, more harmful, agents might enter the semen. The other is that it suggests that these or other agents in the semen not only might affect the male reproductive tract and accessory sex organs but also might gain access to the female reproductive tract or the conceptus (Hales *et al.,* 1986). For example, alcohol has been found to pass rapidly from the bloodstream into prostatic secretion and the semen. However, motility of spermatozoa *in vitro* is unaffected by addition of 2–5% alcohol, suggesting that spermatozoa are relatively resistant to its effects. Agents that have been found in semen following their administration for therapeutic purposes include salicylates, sulfanamides, ampicillin, erythromycin, cephalexin, phenytoin, and tetracycline (see Mann and Lutwak-Mann, 1981). However, they appear not to affect the structural integrity or motility of spermatozoa. In addition, trimethoprim is used to treat bacterial prostatitis and other chronic and acute infections because of its ability to enter the male genital tract (Mann and Lutwak-Mann, 1981). In most cases, the possible effects of these agents on specific reproductive processes have not been closely examined.

Some other drugs that enter the semen have been reported to adversely affect reproductive processes. Thalidomide administered orally to male rabbits was found to enter the semen and to remain for several days. Pregnancy wastage was reported to occur in female rabbits mated with these males. The drug is rapidly accumulated by rabbit spermatozoa *in vitro* and not removed by repeated washing, suggesting that it is sequestered within or tightly bound to spermatozoa. In addition, methadone given orally is rapidly excreted into the semen in rabbits and humans, and male rats given methadone before mating are reported to yield progeny of lower weight at birth and weaning. It also reduces the ejaculate volume by inhibiting secretion of the seminal vesicles and prostate. Orally administered tranexamic acid, a competitive inhibitor of plasminogen activator, enters the semen and inhibits the fibrinolytic activity in seminal plasma.

Some agents have been found to adversely affect specific regions of the male tract or spermatozoa within it. Dibromochloropropane (DBCP), a nematocide that was used to fumigate grain and soil, is metabolized to epichlorohydrin and in turn to α-chlorohydrin. At higher doses, these agents can cause increased capillary permeability in the epididymis leading to vascular damage, sloughing of the epithelium, fibrosis, and sperm granuloma and spermatocoel formation. Epichlorohydrin is used as a solvent in resins, varnish, lacquers, paints, and nail polish. At lower doses, both epichlorohydrin and α-chlorohydrin produce reversible infertility by acting on epididymal spermatozoa to reduce respiratory activity and motility (Waller *et al.*, 1985). It has also been reported to inhibit binding of specific epididymal secretory proteins to the sperm surface (Tsang *et al.*, 1981). Cadmium can also damage the caput epididymidis by causing disorganization of capillary intercellular junctions, and its effects can be prevented by pretreatment with zinc or selenium. In addition, the epididymis is a target for methyl chloride used in the production of silicones, leaded gasoline, and polystyrene. In rats that had inhaled this compound, neutrophils accumulated adjacent to the epididymal epithelium, vacuoles appeared in the epithelium in other areas, and bilateral epididymal granulomas developed later in the tail of the epididymis (Chapin *et al.*, 1984).

VI. Conclusions

The duct system of the male reproductive tract is crucial for ensuring fertility. Not only is it a conduit and reservoir for spermatozoa, but the epididymal portion also serves an obligatory role in the maturational changes that produce motile spermatozoa capable of binding to the zona pellucida. Epididymal function is an active area of research, and it is likely that the specific mechanisms and the molecules involved in some or all of these processes will soon be identified. However, the role of the accessory glands in reproduction is less well understood. Because accessory-gland products are required for *in vivo* fertilization, an important general role of accessory-gland secretions is to help transport spermatozoa into the female reproductive tract and to modify the environment for optimum sperm function and survival. Additional studies are needed to learn the specific roles of individual secretory products of the accessory glands in the reproductive process. At the present time, however, these products (Table I) can serve as valuable indicators of the effects of toxic agents on the male tract.

There is less known about the effects of toxic substances on the duct system and accessory glands of the male reproductive tract than on the testis. However, substances that are toxic to the testis are often toxic to other parts of the male reproductive tract as well. This may be for two reasons. One is that the different parts of the tract may have similar susceptibilities to damage by toxic substances because of similar metabolic characteristics and physiological control mechanisms. The other is that the duct system and accessory glands are highly dependent on androgens, and substances that damage the testis often result in altered androgen production. Thus, before we can understand the mechanisms of action of particular toxic agents, it will be necessary to determine if substances that affect the ducts and accessory glands do so directly or indirectly.

The effects of toxic agents on the ducts and accessory glands have usually been assessed morphologically. This will clearly remain an important method for detecting damage to the male reproductive system. However, additional methods are needed to identify the primary causes that result in the visible effects. Also, subtle metabolic or physiologic changes in the ducts or glands might have profound effects on sperm maturation, motility, or fertilizing ability without producing obvious morphological changes. Other changes, such as perturbation of immunosuppressive activities leading to immune-mediated infertility, could have long-term effects on reproduction that would not be detected by morphological studies. In addition, some toxic substances probably cross the epithelium of the ducts and accessory glands more readily than the seminiferous epithelium because of the "blood–testis" barrier provided by tight junctions between Sertoli cells. Such agents may have dual effects on reproduction, acting rapidly and at low concentrations on the ducts and accessory glands, but also affecting the testis more slowly or at higher concentrations. Effects on the tract and accessory glands might not be detected in morphological studies, even though reproduction is altered.

Because of such considerations, more specific approaches need to be used to monitor the effects of toxic agents on the male reproductive system and to determine the specific sites and mechanisms of action of agents that adversely affect male reproduction. What is known about the nature and role of secretory products of the ducts and glands of the male reproductive tract can be used to develop more specific and quantitative assays for determining the effects of toxic agents on reproductive processes. These findings can in turn be correlated with parameters such as the rate of appearance of toxic substance in the semen, the time course of effect, and the period of recovery in experimental animals. This type of information is needed to provide a basis on which to develop qualitative and quantitative ways to monitor changes in semen composition in humans exposed to potentially toxic environmental agents and to predict their effects on human reproduction. Biochemical end points offer considerable promise for estimating potential risk and for analyzing the mechanisms of action of toxic substances acting on the duct system and accessory glands of the male reproductive system.

References

Aarons, D., Speake, J. L., and Poirier, G. R. (1984). Evidence for a proteinase inhibitor binding component associated with murine spermatozoa. *Biol. Reprod.* **31,** 811–817.

Abrescia, P., Lombardi, G., De Rosa, M., Quagliozzi, L., Guardiola, J., and Metafora, S. (1985). Identification and preliminary characterization of sperm-binding protein in normal human semen. *J. Reprod. Fertil.* **73,** 71–77.

Acott, T. S., Katz, D. F., and Hoskins, D. D. (1983). Movement characteristics of bovine epididymal spermatozoa: Effects of forward motility protein and epididymal maturation. *Biol. Reprod.* **29,** 389–399.

Bedford, J. M. (1975). Maturation, transport, and fate of spermatozoa in the epididymis. *In* "Handbook of Physiology, Section 7: Endocrinology, Male Reproductive System" (D. W. Hamilton and R. O. Greep, eds.), Vol. 5, pp. 303–317. American Physiological Society, Washington, D.C.

Brandes, D. (1974). Fine structure and cytochemistry of male sex accessory organs. *In* "Male Accessory Sex Organs" (D. Brandes, ed.), pp. 18–113. Academic Press, New York.

Brooks, D. E. (1983a). Epididymal functions and their hormonal regulation. *Aust. J. Biol. Sci.* **36,** 205–221.

Brooks, D. E. (1983b). Effect of androgens on protein synthesis and secretion in various regions of the rat epididymis, as analyzed by two-dimensional gel electrophoresis. *Mol. Cell. Endocrinol.* **29,** 255–270.

Carr, D. W., Usselman, M. C., and T. S. Acott (1985). Effects of pH, lactate, and viscoelastic drag on sperm motility: A species comparison. *Biol. Reprod.* **33,** 588–595.

Chapin, R. E., White, R. D., Morgan, K. T., and Bus J. S. (1984). Studies of lesions induced in the testis and epididymis of F-334 rats by inhaled methyl chloride. *Toxicol. Appl. Pharmacol.* **76,** 328–343.

Clavert, A., Cranz, C., and Brun, B. (1981). Epididymal vascularization and microvascularization. *In* "Epididymis and Fertility: Biology and Pathology, Vol. 8, Progress in Reproductive Biology" (C. Bollack and A. Clavert, eds.), pp. 48–57. Karger, Basel.

Cosentino, M. J., Takihara, H., Burhop, J. W., and Cockett, A. T. K. (1984). Regulation of rat caput epididymidis contractility by prostaglandins. *J. Androl.* **5,** 216–222.

Cuasnicu, P. S., Gonzales Echeverria, F., Piazza, A. D., Cameo, M. S., and Blaquier, J. A. (1984). Antibodies against epididymal glycoproteins block fertilizing ability in rat. *J. Reprod. Fertil.* **72,** 467–471.

de Lamirande, E., and Gagnon, C. (1984). Origin of a motility inhibitor within the male reproductive tract. *J. Androl.* **5,** 269–276.

Djakiew, D., Byers, S. W., Lewis, D. M., and Dym, M. (1985). Receptor-mediated endocytosis of α_2-macroglobulin by principal cells in the proximal caput epididymidis *in vivo. J. Androl.* **6,** 190–196.

Eddy, E. M., Vernon, R. B., Muller, C. H., Hahnel, A. C., and Fenderson, B. A. (1985). Immunodissection of sperm surface modifications during epididymal maturation. *Am. J. Anat.* **174,** 225–237.

Fitzgerald, M. J. T. (1978). "Human Embryology: A Regional Approach." Harper and Row, Hagerstown, Maryland.

Flickinger, C. J., Wilson, K. M., and Gray, H. D. (1984). The secretory pathway in the mouse epididymis as shown by electron microscopy radioautography of principal cells exposed to monensin. *Anat. Rec.* **210,** 435–448.

Fournier-Delpech, S., Hamamah, S., Tananis-Anthony, C., Courot, M., and Orgebin-Drist, M.-C. (1984). Hormonal regulation of zona-binding ability and fertilizing ability of rat epididymal spermatozoa. *Gamete Res.* **9,** 21–30.

Gonzales-Echeverria, F., Cuasnicu, P. S., Piazza, A., Pineiro, L., and Balquier, J. A. (1984). Addition of an androgen-free epididymal protein extract increases the ability of immature hamster spermatozoa to fertilize *in vivo* and *in vitro. J. Reprod. Fertil.* **71,** 433–437.

Hales, B. F., Smith, S., and Robaire, B. (1986). Cyclophosphamide in the seminal fluid of treated males: Transmission to females by mating and effect on pregnancy outcome. *Toxicol. Appl. Pharmacol.* **84,** 423–430.

Hamilton, D. W. (1975). Structure and function of the epithelium lining the ductuli efferentes, ductus epididymis, and ductus deferens in the rat. *In* "Handbook of Physiology, Section 7: Endocrinology, Vol. 5, Male Reproductive System" (D. W. Hamilton and R. O. Greep, eds.), pp. 259–301. American Physiological Society, Washington, D.C.

Harris, S., Mansson, P. -E., Tully, D. B., and Burkhard, B. (1983). Seminal vesicle secretion IV gene: Allelic differences due to a series of 20-base-pair direct tandem repeats within an intron. *Proc. Natl. Acad. Sci. U.S.A.* **80,** 6460–6464.

Hermo, L., Clermont, Y., and Morales, C. (1985). Fluid-phase and absorptive cells of the rat ductuli efferentes. *Anat. Rec.* **211,** 285–294.

Isaacs, W., and Coffey, D. S. (1984). The predominant protein of canine seminal plasma is an enzyme. *J. Biol. Chem.* **259,** 11520–11526.

James, K., and Hargreave, T. B. (1984). Immunosuppression by seminal plasma and its possible clinical significance. *Immunol. Today* **5,** 357–363.

Langman, J. (1963). "Medical Embryology." Williams & Wilkins, Baltimore, Maryland.

Mann, T., and Lutwak-Mann, C. (1981). "Male Reproductive Function and Semen." Springer-Verlag, Berlin.

McLachlan, J. A. (1981). Rodent models for perinatal exposure to diethylstilbestrol and their relation to human disease in the male. *In* "Developmental Effects of Diethylstilbestrol (DES) in Pregnancy" (A. L. Herbst and H. A. Bern, eds.), pp. 148–157. Thieme-Stratton, New York.

Muller, B. (1983). Genital tract proteins in the male rabbit: I. Localization of uteroglobin. *Andrologia* **15,** 380–384.

Narbaitz, R. (1974). Embryology, anatomy, and histology of the male sex accessory glands. *In* "Male Accessory Sex Organs" (D. Brandes, ed.), pp. 3–15. Academic Press, New York.

Newbold, R. R., and McLachlan, J. A. (1985). Diethylstilbestrol associated defects in murine genital tract development. *In* "Estrogens in the Environment" (J. A. McLachlan, ed.), pp. 288–315. Elsevier, New York.

Orgebin-Crist, M. -C., and Fournier-Delpech, S. (1982). Sperm–egg interaction. Evidence for maturational changes during epididymal transit. *J. Androl.* **3,** 429–433.

Orgebin-Crist, M. -C., Danzo, B. J., and Davies, J. (1975). Endocrine control of the development and maintenance of sperm fertilizing ability in the epididymis. *In* "Handbook of Physiology, Section 7: Endocrinology, Vol. 5, Male Reproductive System" (D. W. Hamilton and R. O. Greep, eds.), pp. 319–338. American Physiological Society, Washington, D.C.

Ostrow, R. S., Zachow, K. R., Niimura, M., Okagaki, T., Muller, S., Bender, M., and Faras, A. J. (1986). Detection of papillomavirus DNA in human semen. *Science* **231,** 731–733.

Poirier, G. R., and Jackson, J. (1981). Isolation and characterization of two proteinase inhibitors from the male reproductive tract of mice. *Gamete Res.* **4,** 555–569.

Rugh, R. (1968). "The Mouse. Its Reproduction and Development." Burgess, Minneapolis, Minnesota.

Saling, P. M. (1982). Development of the ability to bind to zonae pellucidae during epididymal maturation: Reversible immobilization of mouse spermatozoa by lanthanum. *Biol. Reprod.* **26,** 429–436.

Sujarit, S., and Pholpramool, C. (1985). Enhancement of sperm transport through the rat epididymis after castration. *J. Reprod. Fertil.* **74,** 497–502.

Tash, J. S., Kakar, S. S., and Means, A. R. (1984). Flagellar motility requires the cAMP-dependent phosphorylation of a heat-stable NP-40-soluble 56 kd protein, axokinin. *Cell* **38,** 551–559.

Tsang, A. Y. F., Lee, W. M., and Wong, P. Y. D. (1981). Effects of antifertility drugs on epididymal protein secretion, acquisition of sperm surface proteins and fertility in male rats. *Int. J. Androl.* **4,** 703–712.

Usselman, M. C., and Cone, R. A. (1983). Rat sperm are mechanically immobilized in the cauda epididymis by "immobilin," a high molecular weight glycoprotein. *Biol. Reprod.* **29,** 1241–1253.

Vendrely, E. (1981). Histology of the epididymis in the human adult. *In* "Epididymis and Fertility: Biology and Pathology, Vol. 8, Progress in Reproductive Biology" (C. Bollack and A. Clavert, eds.), pp. 21–33. Karger, Basel.

Waller, D. P., Killinger, J. M., and Zaneveld, L. J. D. (1985). Physiology and toxicology of the male reproductive tract. *In* "Endocrine Toxicology" (J. A. Thomas, K. S. Korach, and J. A. McLachlan, eds.), pp. 269–333. Raven, New York.

Weininger, R. B., Fisher, S., Rifkin, J., and Bedford, J. M. (1982). Experimental studies on the passage of specific IgG to the lumen of the rabbit epididymis. *J. Reprod. Fertil.* **66,** 251–258.

Wilson, E. M., Viskochil, D. H., Bartlett, R. J., Lea, O. A., Noyes, C. M., Petrusz, P., Stafford, D. W., and French, F. S. (1981). Model systems for studies on androgen-dependent gene expression in the rat prostate. *In* "The Prostatic Cell: Structure and Function, Part A. Morphologic, Secretory, and Biochemical Aspects" (G. P. Murphy, A. Sandberg, and J. P. Karr, eds.), Part A, pp. 351–380. Alan R. Liss, New York.

Yeung, C. H., Cooper, T. G., and Waites, G. M. H. (1980). Carnitine transport into the perfused epididymis of the rat: Regional differences, stereospecificity, stimulation by choline, and the effect of other luminal factors. *Biol. Reprod.* **23,** 294–304.

Zaneveld, L. J. D., and Chatterton, R. T. (1982). "Biochemistry of Male Reproduction." Wiley, New York.

Zaneveld, L. J. D., and Tauber, P. F. (1981). Contribution of prostatic fluid components to the ejaculate. *In* "The Prostatic Cell: Structure and Function, Part A. Morphologic, Secretory, and Biochemical Aspects" (G. P. Murphy, A. Sandberg, and J. P. Karr, eds.), pp. 265–277. Alan R. Liss, New York.

Chapter 4

Endocrinology and Paracrinology of the Testis

RICHARD M. SHARPE

MRC Reproductive Biology Unit
Edinburgh, Scotland EH3 9EW

I. Introduction

As testicular growth and function are dependent on a particular pattern of endocrine and paracrine events and because the testis is also an endocrine organ, it may be possible to evaluate and/or predict whether a drug or hormone will have testicular effects by monitoring these markers. To permit such evaluation, it is first essential to understand the "normal" endocrine and paracrine mechanisms involved in the control of testicular development and function, and that is the objective of this chapter. In reviewing what is understood about such processes, particular attention has been given to the predictive value of individual hormones/factors: that is, will their measurement reliably indicate minor or major testicular dysfunction? Most of the information presented is based on studies of the laboratory rat, but in general the findings are probably applicable to all mammals including man.

The growth, development, and maintenance of the mammalian testis are under dual control of the endocrine and paracrine control systems. Endocrine control is exercised via secretion of the pituitary gonadotropins luteinizing hormone (LH) and follicle-stimulating hormone (FSH), and prolactin, and it is through the changing pattern of secretion of these hormones that overall control of testicular function is exercised centrally, as the secretion of these pituitary hormones is under hypothalamic control (see Chapter 5). In addition, other peripherally produced hormones with ubiquitous effects on cells, such as growth hormone and insulin, also exert testicular effects. However, these will not be considered further here other than to point out that disturbances in the production of such hormones as occur, for example, with growth hormone deficiency or diabetes, can result in secondary dysfunction of the testis (for review, see Sharpe, 1982).

In contrast to the gonadotropins, the role of paracrine (local) control mechanisms is to coordinate the various functions of the different testicular cell types and/or to modulate the testicular actions of pituitary gonadotropins according to local conditions and requirements. Such control mechanisms can therefore in some instances be viewed as "fine-tuning" processes, although this description perhaps belies their importance. It should be emphasized that our knowledge of testicular paracrinology is very poor in relation to what is known about testicular endocrinology, primarily because it is a field that has only developed since 1980. However, the potential importance of paracrine control mechanisms appears considerable, and it is now thought that dysfunction of one or more such processes within the testis could account for the majority of the cases of idiopathic infertility in men (see Sharpe, 1986). But perhaps most importantly, it is the disturbance of such local processes by the peripheral administration of drugs or hormones that is the most worrying factor, as such effects are likely to be subtle and subclinical and therefore extremely difficult to detect. The reasons for this are discussed in detail in Section III.

II. Endocrine Control of the Testis

A. Follicle-Stimulating Hormone, the Sertoli Cell, and Spermatogenesis

As has been detailed in Chapter 2, the Sertoli cell is the cell type by which the complex process of spermatogenesis is controlled and synchronized. It is therefore not surprising that the Sertoli cell is a primary target for endocrine and paracrine (Section III) hormones. Other than general metabolic hormones such as insulin, FSH is the only peripheral hormone known to act on the Sertoli cells, but surprisingly little is known about its actions, particularly in the adult testis. Sertoli cells function quite differently in the prepubertal testis than they do in the adult testis, but most of what is known about Sertoli cell function and about the effects of FSH are derived from studies on immature animals. This limitation is seldom admitted, but it is a likely reason for the vast gaps in our understanding of Sertoli cell function in the adult testis, since so much attention has been focused on the immature testis, and such information is not necessarily applicable to the adult.

FSH is the hormone of puberty, and in all mammals so far investigated it is the rising serum level of FSH that triggers testicular growth and that initiates spermatogenesis and expansion of the seminiferous tubules (see Section IV and Fig. 3 for more details). Exactly how FSH induces such major changes is virtually unknown, but once the testis is near to full development, then serum FSH decreases to adult levels and subsequently will only increase again if there is major dysfunction of the Leydig or Sertoli cells (discussed later). During early puberty, several of the functions of the Sertoli cell [androgen-binding protein (ABP) secretion, tubule fluid production and possibly inhibin secretion] that were previously controlled by FSH may become primarily testosterone responsive (see Jegou *et al.,* 1983a). This switch-over is understandable when it is considered that testosterone is the hormone of "maturity" and that its levels are rising as FSH levels are declining (see Section IV and Fig. 3).

1. Follicle-Stimulating Hormone Effects on the Sertoli Cell

FSH has been shown to exert a number of generalized effects on the Sertoli cells, such as to increase size, protein synthesis, and seminiferous tubule fluid production; *in vitro,* FSH can alter the shape and morphology of the Sertoli cell (for details, see Means *et al.,* 1976; Setchell, 1978; Rich and de Kretser, 1983); FSH acts on the Sertoli cell via the classical pathway involving stimulation of cyclic AMP, activation of protein kinase, and protein phosphorylation. More specifically, FSH stimulates the secretion of inhibin and plasminogen activator, the former acting as a long-loop negative feedback regulator of pituitary FSH release (see Chapter 5) and the latter having a possible role in germ-cell translocation (see Section II).

FSH has also been shown to stimulate secretion by the Sertoli cell of androgen-binding protein (ABP), although as testosterone can also perform this function, there is still some controversy as to whether the FSH effects are attributable to LH contamination of the FSH preparation and/or to enhancement of testosterone via the secretion of paracrine factors (see Section III) or the recruitment of new Leydig cells (see Section II,B). This area is of particular interest as it has been suggested that in the past FSH may in some way improve the testicular uptake and/or utilization of testosterone, although the data on which this suggestion was founded is open to question. It is also quite clear that any effect of FSH on ABP secretion would be irrelevant, as the tubular levels of ABP can only account for a small fraction of intratubular testosterone. In fact, the role of ABP is still unknown, although it may involve androgen transport to the epididymis (see Rich and de Kretser, 1983). More recent studies in rats have reopened the issue of FSH effects on androgen utilization by demonstrating quantitative maintenance of spermatogenesis in the presence of only 15% of normal intratesticular testosterone but near-normal FSH levels (Rea *et al.,* 1980). As earlier studies in androgen-treated hypophysectomized rats had failed to show quantitative maintenance of spermatogenesis with comparable levels of androgen, it is concluded that the different results obtained in the most recent study indicate the ability of FSH to improve the utilization of testosterone. The fact that FSH can influence testosterone production by modulating Leydig cell supply and function (discussed below), and other evidence suggesting effects of testosterone on Sertoli cell inhibin production (Ultee-van Gessel *et al.,* 1986), certainly all suggest that there is an intricate interrelationship between FSH, testosterone, and the Sertoli cell.

One of the problems encountered in the study of FSH effects on the Sertoli cell *in vivo* has been that such effects are usually not demonstrable in normal animals, particularly adults, and it is only after removal of the pituitary or through the use of completely immature (< 15 days of age) rats that FSH effects become demonstrable. This is the case for FSH stimulation of both protein synthesis and ABP secretion by the Sertoli cell. The conclusion from such observations is that from early puberty onward, FSH effects on the Sertoli cell are maximal and cannot be further enhanced by exogenous treatment.

2. Predictive Value of the Measurement of Sertoli Cell Function

When there is major disruption of spermatogenesis, there is associated dysfunction of the Sertoli cell, as evidenced by an 80–90% decrease in the secretion of seminiferous tubule fluid, ABP, and inhibin (Au *et al.,* 1983; Jegou *et al.,* 1983b), and as a result of the latter decrease, serum FSH levels rise (see Tables I and II). Measurement of any one of these parameters will therefore have value in indicating whether an administered compound has caused major disruption of spermatogenesis. However, with the exception of serum FSH, all of the other parameters are difficult to measure, particularly on a large scale, and in situations of such major tubular dysfunction there is almost certain to be an associated reduction in testicular weight. As the latter is far easier to measure and is probably of equally good predictive value, it would seem foolish to undertake more laborious and expensive measurements. At the same time, the limitations of simply measuring testicular weight should be kept in mind. First, in rodents such as the rat, major disruption of the seminiferous tubules is often associated with a large increase in the volume of extratubular interstitial fluid (see Table I), which

TABLE I

Changes in Testicular Weight, in Testicular Interstitial Fluid (IF) Volume, and in the IF Levels of a Putative Paracrine Hormone (IF Factor) in Relation to Hormonal Parameters Following the Experimental Induction of Seminiferous Tubule Dysfunction in Adult Rats (Mean ± SD)[a]

Treatment group	Number of rats	Testis weight (mg)	Recovered IF volume (μl/testis)	IF testosterone (ng/ml)	IF factor[b]	Serum hormone level (ng/ml) LH	FSH	Testosterone
Controls	7	1744 ± 157	94 ± 17	345 ± 184	234 ± 57	46 ± 12	300 ± 43	5.0 ± 3.5
Bilaterally cryptorchid (BCD)	4	590 ± 101[e]	144 ± 35	116 ± 65[c]	331 ± 41[c]	157 ± 25[e]	480 ± 43[e]	2.4 ± 0.8
Short-term BCD								
"Scrotal" testis	8	1032 ± 76[e]	220 ± 58[e]	326 ± 239	309 ± 61[c]	112 ± 15[e]	385 ± 68[d]	5.1 ± 6.7
Abdominal testis		546 ± 75[e,f]	110 ± 27[f]	78 ± 70[e,f]	428 ± 80[e,f]			

[a]Results are shown for rats with severe disruption of seminiferous tubule function induced by bilateral cryptorchidism 8 weeks previously and for rats initially made bilaterally cryptorchid but in which one of the testes (the "scrotal" testis) was allowed to redescend into the scrotum after 4–7 days. Note that in this redescended testis the IF level of testosterone returned to normal and there was partial recovery of testicular weight, indicating some recovery of spermatogenesis; however, in these testes there was also a large increase in the testicular IF volume. Note also that the increase in the serum levels of LH and FSH was most pronounced in the animals with two severely damaged testes.

[b]Levels are expressed in terms of the enhancement of hCG-stimulated testosterone production (ng/10^6 cells) when added to Leydig cells *in vitro* (see Sharpe *et al.*, 1986).

[c]$p < 0.05$ versus control.

[d]$p < 0.01$ versus control.

[e]$p < 0.001$ versus control.

[f]$p < 0.001$ compared with the contralateral "scrotal" testis.

will to some extent compensate for the reduction in testicular weight resulting from germ-cell loss. Therefore, testicular weight only provides a qualitative indication of the degree of spermatogenic disruption, and the "true" reduction in testicular weight may be underestimated. This problem can be avoided easily by draining off the interstitial fluid (see Sharpe and Cooper, 1983). Second, changes in Sertoli cell secretory function will generally precede any decline in testicular weight, so that if the time course of effect of administered compounds is of interest, measurement of testicular weight alone may be too imprecise (see Table II). Third, it is possible that there may be a selective effect of a compound on a single aspect of Sertoli cell function that is not reflected by any detectable change in testicular weight; however, as yet there has been no report of such a selective change.

Measurement of serum FSH levels is extremely simple in the rat and human, although it is less easy in nonhuman primates. With certain provisos it is an accurate indicator of Sertoli cell dysfunction, that is, impaired inhibin secretion (Tables I and II). First, as pituitary FSH secretion is under dual negative feedback control from the testis via inhibin and testosterone, any decrease in testosterone secretion as a result of Leydig cell dysfunction may lead to an increase in serum FSH levels (see Chapter 5 and Table II). Second, FSH levels are normally raised during puberty, so it is important to know (a) the timing of puberty in the species under investigation and (b) whether test compounds administered to the animals may have altered the timing of puberty. Third, in men and perhaps other longer lived species, serum FSH (and LH) levels rise during old age, probably as the result of an age-related decrease in testosterone secretion coupled with some degree of "natural" seminiferous tubule damage.

B. Luteinizing Hormone, the Leydig Cell, and Testosterone Secretion

Luteinizing hormone (LH) acts exclusively on testicular Leydig cells and is the primary regulator of testosterone secretion. Testosterone has both intratesticular (i.e., paracrine) actions and a host of peripheral effects on reproductive (secondary sexual organs) and nonreproductive tissues (bone, muscle, skin, liver, central nervous system, etc.), and LH therefore represents the vehicle by which the brain is able to exert overall control of the timing of testosterone production. The levels of testosterone required intratesticularly for the maintenance of spermatogenesis (see Section III) are far higher than those required for maintenance of peripheral androgen effects and for negative feedback effects on pituitary LH secretion (see Fig. 1 for comparison of testosterone levels in the testis, and in spermatic and periph-

TABLE II

Changes in Testicular Weight, in Testicular Interstitial Fluid (IF) Volume, and in the IF Levels of a Putative Paracrine Hormone (IF Factor) in Relation to Hormonal Parameters Following Treatment of Adult Rats with a Single Dose of Either Ethane Dimethanesulfonate (EDS)[a] or Methoxyacetic Acid (MAA)[b] (Mean ± SD)

Treatment group	Number of rats	Testis weight (mg)	Recovered IF volume (μl/testis)	IF testosterone (ng/ml)	IF factor[c]	Serum hormone level (ng/ml)		
						LH	FSH	Testosterone
Controls	8	1562 ± 86	65 ± 13	290 ± 160	474 ± 150	75 ± 51	250 ± 50	6.0 ± 5.1
EDS + 1 day	7	1498 ± 32	84 ± 30	76 ± 43^{e}	310 ± 92	104 ± 72	330 ± 80	1.1 ± 0.4^{e}
EDS + 3 days	6	1480 ± 20	46 ± 12	$<2^{f}$	872 ± 155^{f}	462 ± 329^{d}	485 ± 48^{f}	0.1 ± 0.03^{f}
EDS ± 28 days	10	1179 ± 145^{f}	128 ± 68^{e}	283 ± 123	900 ± 82^{f}	109 ± 66	570 ± 83^{f}	2.4 ± 1.0^{d}
Controls	27	1496 ± 127	75 ± 16	217 ± 119	710 ± 149	37 ± 16	235 ± 49	2.8 ± 1.5
MAA + 1 day	15	1306 ± 46^{d}	54 ± 14^{d}	207 ± 77	928 ± 221^{e}	51 ± 25	276 ± 44^{d}	2.6 ± 1.6
MAA + 3 days	12	1236 ± 50^{e}	57 ± 17^{d}	264 ± 180	767 ± 192	44 ± 19	292 ± 33^{e}	2.7 ± 1.8

[a] Treatment with EDS (75 mg/kg,i.p.) results in total destruction of the Leydig cells within 2–3 days, and the later (28 days) decrease in testicular weight is a consequence of the withdrawal of testosterone; note, however, that intratesticular testosterone levels are back to normal by 28 days due to Leydig cell regeneration (see Bartlett *et al.*, 1986). Note that the increase in FSH levels at 3 days post-EDS is not associated with any change in testicular weight and probably results from the drop in testosterone negative feedback. In contrast, the selective increase in FSH levels at 28 days is clearly indicative of seminiferous tubule dysfunction. Note also that levels of a putative paracrine regulator of the Leydig cells (IF factor) are increased at days 3 and 28.

[b] Treatment with MAA (650 mg/kg, by gavage) results in selective destruction of pachytene spermatocytes only at certain stages of the spermatogenic cycle (see Chapters 2 and 7); in this study approximately 70–75% of seminiferous tubules were damaged by this treatment. Note that compared to the effects of EDS, all changes are relatively minor but that there is a selective increase in FSH levels (and a decrease in testicular weight) at 1 and 3 days after MAA treatment. Note also that there is a transitory rise in levels of the IF factor at 1 day.

[c] Levels are expressed in terms of enhancement of hCG-stimulated testosterone production (ng/10^6 cells) when added to Leydig cells *in vitro* (see Sharpe *et al.*, 1986).

[d] $p < 0.05$ versus control.

[e] $p < 0.01$ versus control.

[f] $p < 0.001$ versus control.

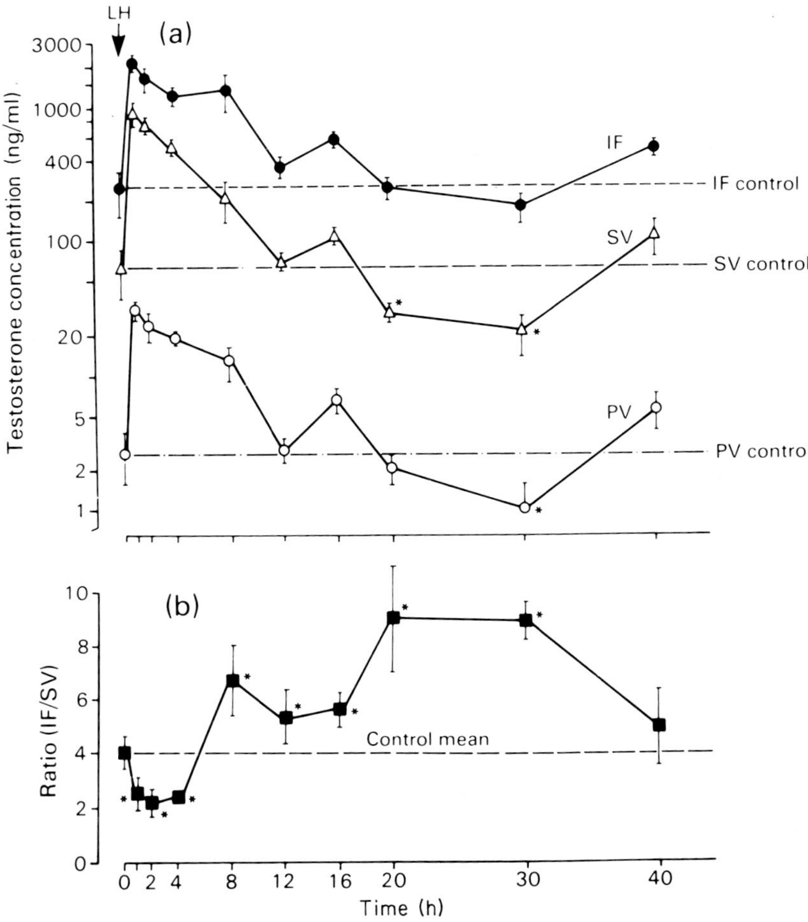

Fig. 1. (a) Comparison of testosterone levels in adult rats, within the testis in testicular interstitial fluid (IF), in spermatic venous blood (SV) and in peripheral venous blood (PV). Values are shown both for untreated control animals (mean ± SD, $N = 6$; time 0) and at intervals of 1–40 h after a single injection of 25 μg LH to provide an artificial "LH pulse" (mean ± SD, $N = 3$ per point). Note that testosterone levels within the testis are always much higher than in the two venous compartments and that the ratio of testosterone levels in IF to the levels in blood draining from the testis in the spermatic vein does not remain constant during the "LH pulse" (b). Note also that testosterone levels are plotted on a logarithmic scale. In (a), asterisks indicate where values fell significantly ($p < 0.02$) below control values; in (b), asterisks indicate where values were significantly ($p < 0.05$ to $p < 0.001$) lower or higher than the corresponding control values. [From Sharpe and Cooper (1983), with permission.]

eral venous blood). This difference means that peripheral administration of synthetic androgens at relatively low doses will support secondary sexual characteristics and libido, while at the same time, suppressing LH secretion and hence lowering the intratesticular levels of testosterone. Therefore, peripheral administration of androgens can lead paradoxically, to suppression of the androgen-dependent process of spermatogenesis. In practice, such suppression is only partial, although severe oligospermia has been induced when LH levels have been suppressed more effectively with a combination of a progestagen and low dose androgen therapy.

1. Luteinizing Hormone Pulses

In all mammals so far studied, including humans, the secretion of LH by the pituitary is pulsatile (see Chapter 5; also see Lincoln and Short, 1980). Changes in both the frequency and the amplitude of the LH pulses are important events during puberty and/or seasonal development of the testis in seasonally breeding mammals (Lincoln and Short, 1980; Wu, 1985). Because LH is secreted in pulses, it follows that Leydig cell testosterone production in response to LH is also pulsatile, each pulse of testosterone being preceded by an LH pulse. An important implication of this arrangement is that single blood samples may give an inaccurate picture of serum LH and testosterone levels, which explains why such measurements are extremely variable. In rats, such variation is equally evident in the serum and intratesticular levels of testosterone. Thus the normal range for serum testosterone in a young adult is 0.5 – 15.0 ng/m, while the levels in testicular interstitial fluid range from 80 to 2500 ng/m (see also Fig. 1 and Table I and II).

2. Estrogens

In the adult rat testis, Leydig cells are also a source of estrogens, although their function is really unknown. When estrogens are administered to animals in small amounts they exert potent antitesticular effects. This is due partly to feedback inhibition of pituitary gonadotropin release but also involves direct antitesticular effects (see Sharpe, 1982). As the Leydig cell is the only cell type with demonstrable estrogen receptors, then it is presumed that such local effects are mediated via disturbance of Leydig cell function. However, although there is plenty of evidence that estrogens (including the estradiol produced by the Leydig cells themselves) can inhibit testosterone production to some degree (see Catt *et al.*, 1980), even total withdrawal of testosterone takes 7 – 14 days before it is manifested by a moderate reduction in testicular weight, whereas the effect of estrogens is more rapid and more pronounced. This indicates that there are considerable gaps in our understanding, and perhaps it will prove that other recently identified Leydig cell products such as β-endorphin and oxytocin (see Section III) play important physiological roles that are antagonized by estrogen exposure.

In immature rats, the Sertoli rather than the Leydig cells are the source of estrogens, and their production is stimulated by FSH (see Sharpe, 1982); at this age, therefore, estrogens may act as a means by which the Sertoli cells can exert negative feedback inhibition of Leydig cell function by aromatizing the small amounts of testosterone produced by the Leydig cells. However, Sertoli cell aromatase activity is lost completely in the rat at about 20 days of age.

3. Control of Leydig Cell Development

It has been generally accepted for many years that Leydig cell numbers and their differentiation are under the control of LH, but this now appears incorrect. LH almost certainly plays a permissive role in Leydig cell development, in that the differentiation of new Leydig cells fails to occur in the complete absence of LH, but it appears that primary control over the Leydig cells is exercised locally via the Sertoli cells. This stems from three pieces of evidence. First, during early puberty FSH treatment (but not LH treatment) of hypophysectomized rats induces the differentiation of new adult-type Leydig cells (Kerr and Sharpe, 1985), and there is growing evidence that the Sertoli cell in the adult testis exerts a controlling influence over the Leydig cells, occurring at least partially via FSH-responsive mechanisms (see Section III,C). Second, local disruption of spermatogenesis, either within a small area of a testis or in one testis but not in the contralateral testis, results in Leydig cell hypertrophy and/or hyperplasia only in the immediate vicinity of the spermatogenic damage (for references and reviews see Sharpe, 1982; de Kretser and Kerr, 1983). Third, and in keeping with the latter findings, when Leydig cells are selectively and completely destroyed by injection of a toxic chemical [ethane dimethane sulfonate (EDS) (see Table II)] repopulation of the testis with Leydig cells subsequently occurs and these cells have a predominantly peritubular origin, and this is particularly striking when there is concomitant disruption of spermatogenesis in the seminiferous tubules (Kerr and Donachie, 1986). It is also relevant that testosterone has been shown to alter inhibin secretion *in vitro* by Sertoli cells and, as this is a selective feedback regulator of FSH secretion, this perhaps provides further indirect evidence for a relationship between FSH stimulation and the intratesticular testosterone supply.

4. Predictive Value of Luteinizing Hormone and Testosterone Measurements

In general, measurements of the serum levels of LH and testosterone are of relatively little predictive value (see Tables I and II). This is particularly true for single blood samples because of LH and testosterone pulsatility (see above), and it is likely that subtle changes in the frequency and/or amplitude of LH pulses, which might be induced by a particular treatment, would go undetected by all but the most detailed investigations, none of which would be feasible in rodents because of their small size. Indeed, apparently normal Leydig cell function (i.e., serum testosterone levels within the normal range) is common in many situations in which overt seminiferous tubule dysfunction has been induced. For example, in adult rats in which severe disruption of the seminiferous epithelium has been induced by experimental bilateral cryptorchidism, serum levels of testosterone are within the normal range and yet, histologically, it is known that there are gross changes in Leydig cell morphology and size (see de Kretser and Kerr, 1983) and intratesticular levels of testosterone are grossly subnormal (see Table I). What is even more difficult to reconcile is that the latter decrease occurs in the face of raised serum levels of LH, and yet when Leydig cells are isolated and their responsiveness to LH is tested *in vitro,* the cells are hyperresponsive (Sharpe *et al.*, 1984). The most acceptable interpretation of such apparently paradoxical findings is that the supply of LH to the Leydig cells *in vivo* is limited by a severe (~ 90%) reduction in testicular blood flow and therefore higher LH levels are needed to maintain "normal" serum testosterone levels. Such a change is sometimes termed compensated Leydig cell failure, but this is probably a misnomer, as the explanation would fail to explain the discrepancy between *in vivo* and *in vitro* data and between the intratesticular and serum levels of testosterone. Indeed, it seems inescapable that some change in the LH–testosterone axis must have occurred, and comparable studies in experimentally cryptorchid rams and other domestic species support this conclusion (see Schanbacher, 1980).

Although I have indicated that measurement of serum LH and testosterone are of little predictive value, such measurements are easy to make and, in conjunction with measurements of FSH and testicular size, they may provide a fuller picture of the endocrine status of the testis (see Tables I and II). However, it would seem preferable to obtain a measurement of the intratesticular levels of testosterone, and while it is certainly more laborious to measure the total testicular testosterone content (as a several-stage extraction/separation process is required), levels of this steroid in testicular interstitial fluid can be measured with ease in the rat with no requirement for extraction. Moreover, interstitial fluid is easy to collect, and the levels of testosterone in this fluid have been shown to accurately reflect both the seminiferous tubule levels and the total testicular content of testosterone (Sharpe and Bartlett, 1985).

C. *Prolactin*

Although it seems clear that prolactin is involved in the normal endocrine regulation of the testis, its role is still poorly understood and is somewhat controversial. Its target cell in the testis is the Leydig cell and, at least in rats and hamsters, there is good evidence that prolactin can regulate the number of Leydig cell LH receptors. Thus, induction of hyperprolactinaemia increases the number of LH receptors and thus the sensitivity of the Leydig cells to LH. However, as the hyperprolactinaemia also causes a severe reduction in the serum levels of LH (and FSH), the testosterone levels remain apparently normal (for review see Sharpe, 1982). The reduction in LH levels also questions whether the increase in the LH receptors induced by the hyperprolactinaemia is solely a Leydig cell response to prolactin or whether it is also a response to the reduced LH levels, as it is clearly established that LH negatively regulates LH-receptor numbers (Sharpe, 1982). Certainly, the situation is not straightforward, as, in men, hyperprolactinemia is clearly associated with impotence and infertility, and severe hyperprolactinemia in rats (induced by a transplantable tumour) is also associated with testicular atrophy.

There is other evidence, again in rodents, to suggest that prolactin may control Leydig cell cholesterol metabolism, either by regulating the number of lipoprotein receptors or by modulating cholesterol esterase activity within the cell (see Catt *et al.*, 1980; Sharpe, 1982).

In view of its uncertain role, measurement of serum prolactin is unlikely to be of predictive value in indicating whether changes in testicular structure and/or function have occurred.

III. Paracrine Control of the Testis

Our understanding of the role of paracrine control processes is that they coordinate the many different functions of the various cell types in the testis in order that the highly complex process of spermatogenesis may proceed in an orderly manner and as efficiently as possible. In considering what is known about such mechanisms (and our information is limited), it is more comprehensible to separately consider paracrine control of the three primary constituent parts of the testis, namely, the vasculature, the seminiferous tubules, and the Leydig cells. In doing so, it should be kept in mind that this division is for the sake of convenience and that it is the overall coordination of the separate functions of these constituent parts that is the key factor (see Fig. 2).

A. Paracrine Control of the Vasculature

The importance of the testicular blood supply has more or less been ignored by the great majority of investigators, despite the fact that the testicular vasculature is unique for several reasons [see Setchell (1978) for full details]. Primarily, it is important because the vast bulk of the testis—the seminiferous tubules, which account for 60–85% of testicular volume—is avascular, and yet this tissue is one of the most highly proliferative epithelia in the body. It is for this reason that the testis is considered to be poised on the brink of hypoxia, and this probably explains why varicoceles and, in particular, testicular torsion can have such catastrophic effects on spermatogenesis

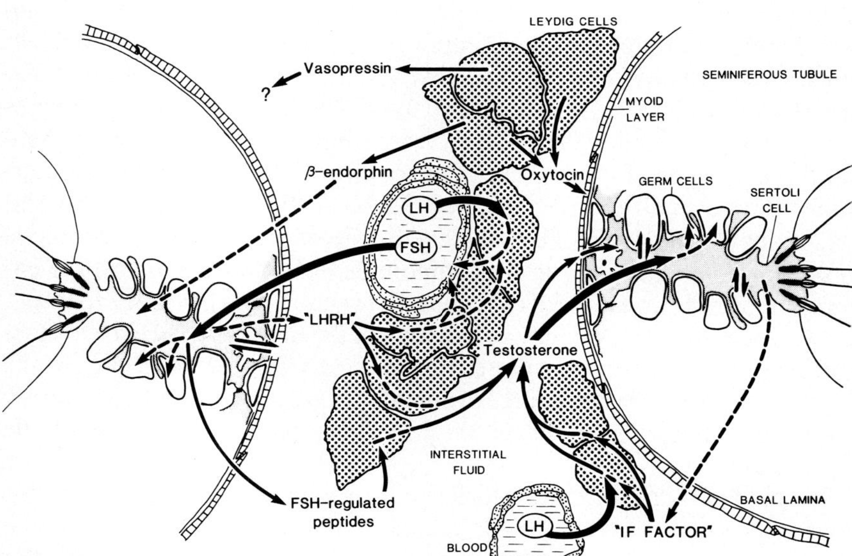

Fig. 2. Diagrammatic summary of the major known paracrine events within the testis and how these interface with gonadotropin action. Note that the many possible paracrine interactions between the Sertoli cell and germ cells are not listed, while at least three separate putative paracrine effects of the Sertoli cells on the Leydig cells are illustrated. It should also be remembered that macrophages, which have been omitted for the sake of clarity, probably also participate in paracrine interactions within the testis.

and fertility (see Setchell, 1978). Another unique feature is that the blood vessels of the testis are unfenestrated, although despite this there is the formation of a copious exudate from the bloodstream — the interstitial fluid — which bathes the Leydig cells and seminiferous tubules and which constitutes the transport medium of the testis (see Chapter 2 on testicular morphology; also see Fawcett *et al.*, 1973). This fluid is present in amounts that would be considered edematous in any other tissue, and in situations of severe seminiferous tubule dysfunction the volume of fluid may be substantially increased, particularly when there is some recovery of testicular weight and tubular function (see Table I). The real importance of the testicular vasculature is that it is the gateway to the testis, and all of the nutritional and endocrine products essential to testicular function can only arrive via the bloodstream (see also Waites and Gladwell, 1982). Therefore, compounds that exert effects on the vasculature are likely to result in major changes in testicular function without actually affecting the levels of nutrients and hormones in the bloodstream. The classic example of this is cadmium, which, via effects on blood vessels in the spermatic cord, results in almost complete testicular atrophy (Setchell, 1978).

Because of its importance, it is not surprising that the testicular vasculature is under some degree of local control, although it must be admitted that our knowledge is very sketchy. The Leydig cells appear to be primary regulators of the vasculature, as LH action on these cells can lead to major increases in vascular permeability, that is, it increases the rate of formation of interstitial fluid, and therefore the rate of entry of nutrients and endocrine hormones; some small changes in testicular blood flow may also occur (for details, see Sharpe, 1983). The Leydig cells are certainly known to be a source of several potentially vasoactive compounds, including angiotensin, β-endorphin, prostaglandins, and estrogens, but present data has failed to implicate any of these compounds as mediators of the LH-regulated effects. Macrophages, which are always present in the testis in close proximity to the Leydig cells, are also a rich source of potentially vasoactive agents, and future studies are likely to implicate these cells in paracrine control of the vasculature.

It would perhaps seem most logical if the seminiferous tubules were able to exert local control of the vasculature, as it is this tissue that is avascular and is the major user of the blood-borne products. Certainly, the rate of testicular blood flow does appear to go hand in hand with spermatogenesis in that reductions in blood flow become progressively more marked as spermatogenic disruption increases (see Wang *et al.*, 1983). However, it is unknown whether this is a cause or an effect of the spermatogenic disruption. Permeability of testicular blood vessels can be altered as a consequence of the local action of LH-releasing hormone (LHRH) on the Leydig cells, and the significance of this is that the Sertoli cells are probably the source of an "LHRH-like" peptide that acts as a coordinator of Sertoli–Leydig cell functions (see below and Fig. 2; also see Sharpe, 1984a,b). The action of LHRH is to reduce vascular permeability (i.e., the opposite effect to LH), but, interestingly, in the presence of raised LH levels LHRH acts synergistically with the LH to increase vascular permeability more quickly than would the LH on its own. LHRH also has small stimulatory effects on testicular blood flow. It seems highly likely that other products of the Sertoli cell will subsequently prove to have vasoactive effects.

Finally, it should be kept in mind that although the testis may exert local control of its vasculature, many other "peripheral" factors can also exert modulatory effects, and these include posture, temperature, altitude (oxygen availability), and peripherally derived vasoactive agents.

Because of our limited understanding of paracrine control of the testicular vasculature, its predictive value is unknown. Measurement of testicular blood flow is not straightforward, and certainly cannot be applied in a routine way to large numbers of animals. As has already been mentioned, reductions in testicular blood flow appear to go hand in hand with reductions in testicular volume, so any change in the latter will infer a commensurate decrease in blood flow. Methods also exist for the measurement of testicular vascular permeability, but these are time consuming and their predictive value is totally unknown. As changes in vascular permeability appear to be reflected by changes in the volume of interstitial fluid that is recovered from the testis (Sharpe and Cooper, 1983), it would seem preferable to use this measure as an easy if crude index of the status of vascular permeability.

B. Paracrine Control of the Seminiferous Tubules and Spermatogenesis

The process of spermatogenesis is extremely complex yet highly organized, and it is for this reason, coupled with the need for precise timing of many sequential events (mitosis, meiosis, germ-cell translocation, spermiation—see Chapter 2 for details) during spermatogenesis, that the need arises for paracrine control mechanisms. Our knowledge of such mechanisms is poor and lacking in detail, and although our understanding has grown rapidly since 1980, the reader should keep in mind that what we currently know must represent the tip of the iceberg. Probably all of the paracrine events within the testis are ultimately directed toward the maintenance of spermatogenesis, but for the sake of clarity they can be broken down into three groups (see also Fig. 2). First are those factors that have an extratubular source but an intratubular site of action. Second are those factors that have an intratubular source and action. Third are those factors that have an intratubular source but an extratubular site of action. This last category comprises mainly factors that act on the Leydig cells, and these are considered below under a separate heading. Before proceeding further, the reader is strongly recommended to read Chapter 2, which describes the organization of spermatogenesis.

1. Paracrine Factors with an Extratubular Source but an Intratubular Action

This currently comprises three hormone groups: testosterone, β-endorphin and related peptides, and oxytocin/vasopressin, all of which are produced by the Leydig cells.

a. Testosterone. Testosterone can perhaps be considered as *the* primary paracrine hormone of the testis, even though it is not exclusively paracrine in that it is also released into the bloodstream to exert secondary peripheral effects. However, within the testis it is the one locally produced hormone on which spermatogenesis depends completely, and in its absence spermatogenesis cannot be maintained. For a hormone with such an all-important role, it is shameful to have to admit that we know absolutely nothing about how testosterone drives spermatogenesis, except that this drive is probably exerted via the Sertoli cells. All that is clear is that testosterone can stimulate the Sertoli cells to secrete ABP and seminiferous tubule fluid, but neither of these are believed to be involved in the actual process of spermatogenesis (see Sharpe, 1983, 1986).

While we know nothing about how testosterone acts, we do know that its actions are directed preferentially toward certain stages of the spermatogenic cycle. Thus, when the secretion of testosterone is inhibited or its action on the Sertoli cells is blocked, then degeneration of the postmeiotic germ cells at stages VII and VIII of the 14-stage rat spermatogenic cycle occurs (Russell *et al.*, 1981; reviewed in Sharpe, 1983). However, while this is true for perhaps the initial 7 – 10 days of testosterone withdrawal, subsequently there may be a more widespread degeneration of germ cells (e.g., Bartlett *et al.*, 1986). It is not known whether this is a specific consequence of androgen withdrawal or simply a "knock-on" effect due to disruption of the orderly spermatogenic process (see below).

Finally, there is exciting recent evidence that shows that testosterone acts on the peritubular myoid cells to stimulate the production of a macromolecule, which then acts on the Sertoli cells to modulate ABP and transferrin secretion (Skinner and Fritz, 1986). This raises the possibility that some of the actions of testosterone on the seminiferous tubule may be mediated by the peritubular cells (see also oxytocin and vasopressin, below).

b. Opiates. β-Endorphin and other proopiomelanocortin (POMC) derived peptides such as α-melanocyte-stimulating hormone are now known to be produced locally by the Leydig cells in at least half a dozen species, including the rat and human (Bardin *et al.*, 1984). Unfortunately, as yet we know virtually nothing about the paracrine role of these hormones, although it is fairly certain that the site of action of β-endorphin is the Sertoli cell. There is some evidence that treatment of neonatal rats with opiate antagonists results in enhancement of testicular size, which suggests that β-endorphin may normally exert inhibitory effects on the Sertoli cells. This is of considerable interest, as there is growing evidence for paradoxical inhibitory effects of LH [and human chorionic gonadotrophin (hCG)] on spermatogenesis in situations in which the supply of testosterone has been maintained; such findings suggest that the Leydig cells are the source of some other factor with potent inhibitory effects on the tubules (for details, see Kerr and Sharpe, 1986), and β-endorphin is an obvious possibility.

c. Oxytocin and Vasopressin. Oxytocin is produced by the Leydig cells in all species so far studied, and this peptide acts on the myoid cells, that surround the seminiferous tubules to bring about their contraction (Wathes, 1984). So far, two distinct types of contraction have been discerned (Worley *et al.*, 1985). The first involves large areas of the seminiferous tubule and visibly propels the luminal contents—this type of contraction is probably therefore involved in the transport of the shed sperm to the rete testis. The second type of contraction is localized and is described as a "dimpling" of the tubular surface. Its function is unknown, but it is possible that it may aid in the translocation of premeiotic germ cells away from the basement membrane and/or aid in the release of sperm at spermiation. Such details should be resolved within the next few years. As yet, little is known about the secretion of Leydig cell oxytocin or whether it is under endocrine or paracrine control, but new evidence suggests that testosterone may be involved.

Vasopressin or a vasopressin-like peptide also appears to be produced in the testis, and in cultures of mixed testicular cells it can cause inhibition of Leydig cell testosterone secretion. However, no such effects are evident in short-term (24-h) cultures of purified Leydig cells, which suggests that vasopressin may first have to act on some other (unidentified) testicular cell type to promote the release of an inhibitor of Leydig cell steroidogenesis. Future studies with cocultures of defined testicular cell types should resolve this question.

2. Paracrine Factors with an Intratubular Source and Action

The process of spermatogenesis involves a precisely timed sequence of events comprising a series of mitotic divisions (six in the rat), meiosis, differentiation, and translocation of the germ cells within the seminiferous epithelium (see Chapter 2). The triggers for such changes are highly selective, in that only certain germs cells at specific stages of the spermatogenic cycle are affected by a given stimulus. Such local specificity must depend partly or wholely on the production of paracrine factors, which presumably are produced by the Sertoli cell and act on the requisite germ cells. At present, we have very little information on the nature and identity of such factors, with one or two exceptions.

A Sertoli cell mitogen—referred to as the seminiferous growth factor (SGF)—has been identified in a wide range of species, including the rat and human, and it is thought that this peptide may be the paracrine regulator of germ-cell mitoses (Bellvé and Feig, 1984). However, little is known about the details of secretion of SGF and, particularly, whether it is produced *only* at the spermatogenic stages when germ-cell mitoses occur (stages IX, XII, XIV, I, IV, and VI in sequence in the rat). However, as SGF has been purified and monoclonal antibodies to SGF have recently been raised, such questions should soon be answered. A further intriguing possibility has also arisen with publication of the full sequence of inhibin (Mason *et al.*, 1985), which has shown homology with transforming growth factor-β (TGF-β). This may mean that in addition to its endocrine role as a feedback regulator of FSH secretion, inhibin (or the TGF-β portion of the molecule) may have paracrine effects on testicular cell growth.

The second fundamental process that occurs during spermatogenesis is meiosis, and to initiate this event the Sertoli cell faces a number of strategic problems. First, meiosis has to commence precisely at stage VII of the spermatogenic cycle, and second, only preleptone spermatocytes must be triggered to enter meiosis, with the other diploid germ cells (the spermatogonia) associated with the Sertoli cell at this stage remaining unaffected. How the latter is achieved is unknown, but there is clear evidence for both the rat and human that two (paracrine) factors with opposing actions—a meiosis-inducing and a meiosis-preventing factor—are produced by the Sertoli cell, and preliminary evidence for the rat suggests that differential secretion of these two peptides at different stages of the spermatogenic cycle is the means by which initiation of meiosis may be timed (for references see Parvinen, 1982; Parvinen *et al.*, 1986).

Two more-or-less coincident major restructuring events occur during stages VII and VIII of the rat spermatogenic cycle—spermiation, whereby the fully differentiated spermatozoa are released into the luminal fluid, and translocation of preleptotene spermatocytes from the basal to the "protected" adluminal compartment behind the inter-Sertoli cell tight junctions (Russell, 1980). At the same time there is a massive increase in the Sertoli-cell secretion of a protease, plasminogen activator (Parvinen, 1982), which in other tissues is associated with tissue restructuring. If this association proves to be causal, then perhaps plasminogen activator can be considered as a paracrine messenger.

The number of morphological and biochemical changes to germ cells during spermatogenesis is vast, and yet we know virtually nothing about how such changes are timed and effected. At least some must involve important paracrine factors, and we know, for example, that the pattern of protein secretion by the Sertoli cell changes considerably during the spermatogenic cycle. In particular, a protein termed simply "cyclic protein 2" is secreted in much higher amounts at stage VI of the spermatogenic cycle than at other stages, and several other proteins show less marked stage-dependent differences in secretion (Wright *et al.*, 1983). Unfortunately, the role of cyclic protein 2 is completely unknown. However, two other proteins of known function also appear to be produced in a cyclical manner by the Sertoli cell. Transferrin, an iron-transporting protein, is produced in greater quantities at stages IX–XIV of the cycle than at other stages, while ceruloplasmin, a copper-transporting protein, may be produced in greater quantities at stage VI (Wright *et al.*, 1983). While the significance of such cyclical changes is unknown, these two proteins are almost certainly produced for the benefit of the postmeiotic germ cells, which, because of their sequestration behind the inter–Sertoli cell tight junctions, do not have access to plasma-derived transferrin and ceruloplasmin, which are presumably present in the interstitial fluid surrounding the tubules.

Whether transferrin and ceruloplasmin can be considered as paracrine hormones or simply as metabolic requirements for the germ cells is open to debate, and there are two other compounds which fall into a similar category: lactate and pyruvate. These are produced by the Sertoli cell probably in response to FSH and are the energy substrates for the postmeiotic germ cells, which are unable to utilize glucose as a source of energy (see Jutte *et al.*, 1983). It is unknown whether the secretion of such products is cyclical, but if FSH is the primary stimulator of production in the adult, as it is in the immature Sertoli cell, then it is relevant that the number of Sertoli-cell FSH receptors and Sertoli-cell responsiveness to FSH show marked changes during the spermatogenic cycle (see Parvinen, 1982).

a. Germ-Cell Effects on the Sertoli Cells. In considering the paracrine control of germ-cell development by the Sertoli cells, no mention has been made of how the Sertoli cell knows when to initiate a given event. Do the germ cells act as the trigger by exerting paracrine effects on the Sertoli cell? There is no direct information on this subject for the adult testis, but studies with immature rat Sertoli cells in culture suggest that addition of different germ-cell types can modify Sertoli cell function, as indicated by ABP secretion (e.g., Galdieri *et al.*, 1984). For such effects to occur, the germ cells have to attach to the Sertoli cells, which suggests that the modifying influence of the germ cells is exercised through the transfer of factors via cell junctions. Less direct evidence from the adult strongly supports the idea that it is the germ cells that "program" the function of the Sertoli cell. Thus, there is a rapid decrease in secretion of a number of Sertoli cell products (seminiferous tubule fluid, ABP, inhibin) following partial (Jégou *et al.*, 1984) or major depletion (Au *et al.*, 1983; Jégou *et al.*, 1983b) of germ cells. More specifically, when one of the four or five layers of germ cells that are normally associated with the Sertoli cell is removed (by X-ray irradiation) and the stage-dependent secretion of plasminogen activator (PA) by tubule fragments is measured, it has been found that *only* removal of the layer of pachytene spermatocytes prevents the peak in PA secretion that normally occurs at stages VII–VIII (see above; also see Vihko *et al.*, 1984; Parvinen *et al.*, 1986).

From such data, it is considered that at a given stage of the spermatogenic cycle, the various functions of the Sertoli cell are dictated by the particular (and unique) germ-cell complement at that stage. It follows that when one or more germ-cell types are missing, then the function of the Sertoli cell may change, and such changes may further affect the development of the germ cells that still remain in contact with the Sertoli cell. The implications of this conclusion are considerable in that they suggest that a drug or hormone that has relatively minor effects on particular germ cells may, secondarily, affect the development of the remaining germ cells. This supposition has been shown to be true for at least one chemical (e.g., 1,3-dinitrobenzene; see Chapter 2). From the toxicological point of view this has serious implications, which are discussed at length at the end of this section.

C. Paracrine Control of the Leydig Cells

The absolute dependence of spermatogenesis on testosterone produced by the Leydig cells has been discussed already. Because of this dependence, it seems logical that the Sertoli cell (the primary site of testosterone action) should exert feedback control of Leydig cell testosterone production to ensure the optimum supply. In considering this paracrine control, it should be kept firmly in mind that the major stimulator of testosterone secretion is endocrine in nature—pituitary LH—and that the most likely role for paracrine agents is to modulate responsiveness of the Leydig cells to LH. As will be seen this may be achieved via one or more different means (see Fig. 2).

1. Morphological Evidence

Disruption of normal Sertoli cell function in the adult rat testis by inducing germ-cell loss by means as diverse as drug treatment, vitamin A deficiency, or temperature elevation results in hypertrophy and/or hyperplasia of the Leydig cells only in the immediate vicinity of the "damage" (see Section II; also see Sharpe, 1982, 1983). A similar phenomenon is demonstrable under physiological conditions in the rat, as comparison of the size of peritubular Leydig cells has shown that those cells in close promixity to seminiferous tubules at stages VII–VIII of the spermatogenic cycle (i.e., the acutely androgen-dependent part of the cycle; see above) are significantly larger than peritubular Leydig cells at other stages of the cycle or perivascular Leydig cells (Bergh, 1983). To induce such changes, the Sertoli cells would have to secrete one or more paracrine factors that acted on the neighboring Leydig cells, and the evidence to this effect is detailed below.

2. Follicle Stimulating Hormone-Regulated Paracrine Control

Treatment of immature hypophysectomized rats with FSH (but not LH) induces multiplication and differentiation of functionally competent adult-type Leydig cells, an effect that must be mediated via the secretion by the Sertoli cell of one or more (paracrine) factors (see Section II; also see Kerr and Sharpe, 1985). The fact that the development of Leydig cells occurs during normal puberty in the face of very high levels of FSH and minimal levels of LH suggests that this phenomenon is of physiological significance (see Section IV). More recently, it has been demonstrated *in vitro* for both the rat and the pig that immature Sertoli cells produce factors in response to FSH that enhance the responsiveness to LH of Leydig cells in culture (e.g., Benahmed *et al.*, 1984; Janecki *et al.*, 1985). The importance of these effects and the *in vivo* effects described above is that in each instance the FSH had little or no effect itself on Leydig cell steroid production but enabled the Leydig cells to produce more testosterone when stimulated with LH.

3. Testicular Luteinizing Hormone—Releasing Hormone

In the rat at least, the Sertoli cells appear to produce an LHRH-like peptide that acts exclusively on the Leydig cells, as only these cells have LHRH receptors. Although "testicular LHRH" has so far not been purified, we already know its probable actions due to studies on the effects of LHRH and LHRH analogs on the testis (for full details see Sharpe, 1984a). LHRH can exert both stimulatory and inhibitory effects on Leydig cell testosterone secretion, the former occurring in the short term (up to 24 h) and the latter occurring during more prolonged (3 days and more) exposure to LHRH, particularly in the presence of concomitant exposure to supranormal levels of gonadotropins. Indeed, it is clear that LHRH action is modulated by the degree of exposure of the Leydig cells to LH, and while it is most likely that the physiological role of "testicular LHRH" is to stimulate testosterone secretion, the gross inhibition of testicular function that can occur after chronic treatment with LHRH analogs has been investigated as a possible contraceptive approach (see Swerdloff and Bhasin, 1984). Whether a comparable LHRH-like factor is present in the human testis is unclear and somewhat controversial, partly because of technical problems and partly because of the obvious difficulties in obtaining normal human tissue (for more details and full references, see Sharpe 1984b).

Perhaps the most important point to make about "testicular LHRH" is that it has the potential to control the interstitial fluid levels of testosterone via at least three separate mechanisms: (1) by directly altering Leydig cell testosterone secretion, (2) by altering vascular permeability, and (3) by altering testicular blood flow; points (2) and (3) were discussed in Section III,A. As interstitial fluid bathes the outside of the seminiferous tubules, then it is the levels of testosterone in this fluid that determine the testosterone levels inside of the tubules (Sharpe and Bartlett, 1985). By regulating the three major processes that affect the levels of testosterone in testicular interstitial fluid, it is thus evident that "testicular LHRH" has all of the correct credentials for a Sertoli cell feedback regulator of the local testosterone supply.

4. Other Factors

Most recently, testicular interstitial fluid from the rat has been shown to contain a nonsteroidal polypeptide factor (or factors) of > 30,000 Da which can greatly enhance Leydig cell testosterone production *in vitro* (Sharpe and Cooper, 1984). Again, this factor appears to be mainly a modulator of LH (and LHRH) action on the Leydig cells, as its effects *in vitro* in the absence of LH are of much smaller magnitude than those that occur in its presence. However, the two most exciting aspects about this factor are (1) that it is produced *in vivo* in normal rats and (2) that the levels of this factor in interstitial fluid can be altered considerably by modulation of testicular function and the endocrine environment (see Sharpe *et al.*, 1986). In essence, any treatment that lowers the intratesticular levels of testosterone (thus posing a potential threat to spermatogenesis) and/or that induces seminiferous tubule dysfunction results in elevation of the levels of this factor (see Tables I and II). In contrast, treatment with LH or hCG to raise intratesticular levels of testosterone to supranormal levels results in a large decrease in the levels of the factor in interstitial fluid. As such changes are precisely what would be predicted for a local regulator of the testosterone supply, it is presumed that this factor (which is clearly not "testicular LHRH") is of physiological importance, although as yet it has not been proved that it emanates from the Sertoli cell.

In keeping with the above findings, there are several reports of isolated seminiferous tubules and/or Sertoli cells releasing factors that are able to enhance Leydig cell testosterone production in response to LH (for references see Sharpe *et al.*, 1986), and at least some of the reports have shown these factors to be under FSH control (see above). As yet none of these factors have been characterized, so it is not possible to decide whether they represent a single factor or several, nor whether they have any similarity to the interstitial fluid factor described above. However, in most instances it has been established that the factors are not LHRH-like.

Finally, mention must be made of an aromatase inhibitor that is produced *in vitro* by isolated seminiferous tubules and that is preferentially secreted at stages VII–VIII of the rat spermatogenic cycle (Boitani *et al.*, 1981). As this is the androgen-dependent period of the cycle and as the Leydig cells are the sole testicular source of estrogens in the adult rat, it is possible that this factor represents yet another way in which the Sertoli cells can optimize their local testosterone supply by ensuring minimal conversion to estrogens by the neighboring Leydig cells.

D. Paracrine Effects Involving Other Cell Types

There is clear evidence that other cell types in the testis may also participate in paracrine control processes. For example, peritubular myoid cells can exert major changes on Sertoli cell structure and function in culture, and such effects are mediated via one or more secretory products (see Skinner and Fritz, 1986). It has already been mentioned that testosterone from the Leydig cells can act on the peritubular cells, which, in response, then modulate Sertoli cell function and also that oxytocin from the Leydig cells acts on the myoid cells (see Fig. 2). Thus, it is clear that there are multiple interactions between each of these different cell types.

Macrophages are a numerically important constituent of the testicular interstitium in the rat, there being one macrophage for every four or five Leydig cells. As the macrophages send out processes that interdigitate between the Leydig cells, and because the former are a rich source of many active peptides, it seems certain that these cells also have an important paracrine role to play. In this respect there is recent evidence that these macrophages may be FSH responsive and may modulate Leydig cell testosterone production (see Yee and Hutson, 1985). However, these data are somewhat controversial and remain to be confirmed.

In general, our knowledge of intratesticular paracrine control processes is insufficient to enable its use as an accurate predictor of intratesticular changes following drug/hormone treatment. However, our methods are improving rapidly and there is every prospect that within the next few years certain paracrine hormones will become established as accurate markers of subtle intratesticular changes. The predictive value of testosterone measurement was discussed in Section II.

In the context of toxicological evaluation, one very important factor arises relating to paracrine events within the testis. We now know that at least eight hormones (oxytocin, vasopressin, LHRH, TRH, CRF, calcitonin, angiotensin, β-endorphin) are produced within the testis (for references see Sharpe, 1984b), which, until the last few years, were considered to be exclusively peripherally or centrally acting. The list is obviously incomplete, and other such hormones will subsequently prove to be present within the testis. The implications of this are considerable, because the peripheral administration of any hormone agonist/antagonist may exert intratesticular effects of which we are currently unaware. Indeed, the lesson was learned with LHRH, because it was only when more potent analogs of LHRH were developed that direct effects on the testis (and ovary) became apparent—yet it must be presumed that many previous experiments involving the peripheral injection of LHRH will have resulted in major or minor direct changes in testicular function. The fact that LHRH effects on the testis are similar to those of LH (the secretion of which is LHRH regulated), in that both modulate testosterone production, simply means that it is impossible to unravel what proportion of any change was LHRH regulated and what proportion was LH regulated.

IV. Endocrinology of Puberty

All of the details given so far in this chapter refer to the adult mammal, but in many ways the situation is different in prepuberty, and particularly during puberty when an orderly sequence of endocrine (and presumably paracrine) events must take place. It is not intended to describe puberty in detail here, and the reader is referred instead to an appropriate review (Swerdloff and Heber, 1981). Only the general and important differences from the adult are outlined in so far as they are likely to influence susceptibility to drug/hormone treatment.

During puberty, the hormonal environment is very different from the adult, with high serum levels of FSH, low serum levels of LH (see Fig. 3), and probably with somewhat higher serum levels of prolactin (see Sharpe, 1982). Comparable changes to those seen in the rat appear to occur in all species, including humans (see Kelch *et al.*, 1983), except that the timing of the changes, (i.e., their onset and duration) is obviously different for each spe-

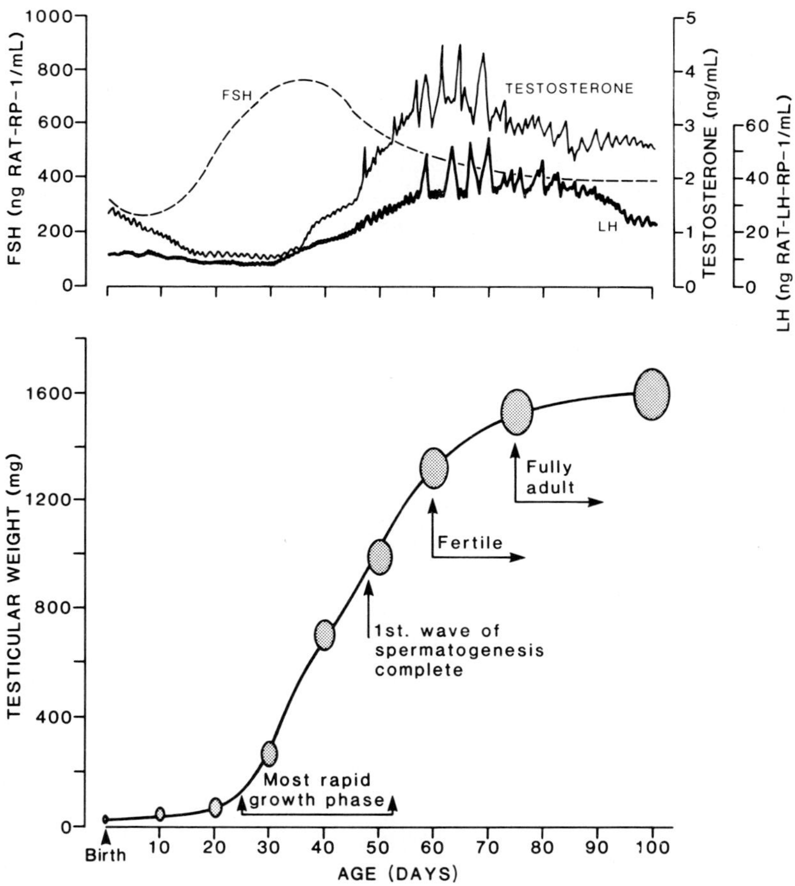

Fig. 3. Endocrine changes during puberty in the rat in relation to testicular growth, spermatogenesis, and fertility. Note that FSH levels are high, particularly in relation to LH levels, for much of puberty but decline in late puberty and adulthood. Levels of LH and testosterone are always extremely variable due to LH pulsatility, but in general their levels rise from mid to late puberty and slowly decline in adulthood. The illustration represents a compilation and consensus of many reports cited in the literature (for references, see Sharpe, 1982) and is intended as a general rather than a precise guide; in particular, the timing and magnitude of the hormonal changes may vary between different strains of rat.

cies. Moreover, hormone levels and testicular structure and function are changing constantly during puberty, and there may also be large differences, for example, in the serum levels of LH and prolactin and in LH pulsatility at night compared with the daytime. There may therefore be a different susceptibility to drugs/hormones during puberty, and perhaps at different stages of puberty, and the following factors may be of particular importance in this respect.

1. The testis is growing rapidly, and there is therefore likely to be relatively high testicular blood flow; also, the nature and distribution of the testicular vasculature is changing (see Setchell, 1978).
2. The germ-cell complement of the testis is changing as spermatogenesis becomes progressively more advanced, and the access of administered compounds to some of the germ cells will change with the formation of the inter–Sertoli cell tight junctions in early puberty.
3. In mid to late puberty, the Leydig cells are becoming mature and are secreting progressively larger amounts of testosterone for the first time since the fetal or early postnatal period.

V. Endocrinology of Male Infertility

It is not possible in this chapter to give a detailed description of the endocrinology and treatment of male infertility, and for details the reader is referred to excellent recent reviews (Hargreave, 1983; Wu, 1985). In the context of the present chapter, the importance of male infertility is that knowledge of the endocrine and paracrine changes in subfertile/infertile men should aid in the interpretation and prediction of the effects of administered compounds on the testis. Before briefly considering the main subgroups of infertile men, it is first appropriate to make some more general observations.

Subfertility in men is common and is generally considered to affect at least 5% of the population. As will be described later, in perhaps 75% of such men the condition is idiopathic and is untreatable. The vast majority of such men produce either normal or subnormal (oligospermia) numbers of sperm, and such sperm may have no obvious morphological or functional abnormalities or they may have single or multiple abnormalities. It is thus obvious that subfertile men are a heterogeneous group, and that their subfertility must stem from many different causes. At present, with the exception of the few treatable cases (see below), there is little that can be done for such patients other than to "characterize" their infertility by hormone measurements, by detailed analysis of sperm numbers, morphology, and motility, and by functional assessment using tests such as the penetration of cervical mucus or the penetration of zona-free hamster oocytes (see Aitken, 1983; Wu, 1985). Valuable as such studies are, they have not identified the initial cause(s) of subfertility. However, as major endocrine changes in such patients are extremely rare, it is likely that intratesticular paracrine defects may be the primary cause of infertility in such men (see Sharpe, 1986).

Recent studies in which damage to selective germ cells in the rat testis has been induced by treatment with industrial chemicals indicates that minor defects in the germ cell complement may give rise to subfertility characterized by a normal hormone profile, only a moderate reduction in sperm count, but a large reduction in fertility (see Chapters 6, 7, and 8). The parallel to much of idiopathic infertility in men is striking, and such chemical-induced effects may prove to be extremely useful tools for improving our understanding of the possible aetiology of infertility in man. Also, as such effects are induced by the administration of industrial chemicals, many of which may be present in the working or home environment of a proportion of the population, it must be kept in mind that such chemicals may even be responsible for some of the cases of male infertility.

As yet, there is no information on paracrine changes in the testes of infertile men, and in the context of this chapter, division of infertile men into subgroups according to their endocrine profiles is the most appropriate.

1. Hypogonadotropic hypogonadism. This accounts for less than 1% of infertile men and because of the deficiency in LH and FSH secretion often presents because of delayed puberty. In the adult it may result from head injury, pituitary tumors, or meningitis, but it is very rare. It is easy to diagnose and in many cases can be treated by LHRH or gonadotropin therapy.
2. Hyperprolactinaemia. This is usually more a problem of loss of libido and erectile impotence, but variable disturbance of spermatogenesis may also occur. As in animals, the raised levels of prolactin are probably associated with altered gonadotropin release. Hyperprolactinaemia in men is usually caused by a pituitary macroadenoma, which is responsive to dopamine agonists but may also require surgery and/or radiotherapy. Control of prolactin levels is associated with the recovery of normal sexual function, but it is possible that spermatogenesis may not revert completely to normal.
3. Overt testicular dysfunction. In men in whom there is gross disruption of spermatogenesis, there are, as in animals, raised serum levels of FSH. Although it may be possible to identify the likely original cause of this dysfunction (e.g., past history of cryptorchidism, mumps orchitis, testicular torsion, etc.), the presence of raised serum FSH levels in an adult is now generally taken as an indication for no further treatment, as the prognosis is poor. Most of such men are azoospermic or severely oligospermic. As in animals, raised serum FSH is invariably associated with small testes in humans.
4. Idiopathic infertility. In the vast majority of infertile men, the endocrine profile is normal, and although there may be varying degrees of oligospermia, in many individuals the semen profile is within a de-

fined normal range. However, it should be noted that a low sperm count is not necessarily indicative of infertility, and, for example, in hypogonadotropic patients treated with gonadotropin-releasing hormone (GnRH) or gonadotropins, sperm counts are invariably low but such men are often fertile (see Wu, 1985). There have been suggestions that there may be subtle deficiencies in the intratesticular androgen supply, but this evidence is not generally applicable, and androgen or hCG therapy of such men has never been proved to be beneficial.

VI. Summary and Future Prospects

This chapter has introduced the reader to the complexities of the hormonal control of the testis. This involves two systems, a relatively straightforward endocrine input that is modulated intratesticularly by a complex paracrine control system (see Fig. 2). Our current interpretation of this arrangement is that overall control of testicular function is exercised via the endocrine input, while the paracrine aspect modulates the endocrine drive according to local requirements and coordinates each of the many intratesticular events that contribute to the process of spermatogenesis. I have emphasized our overall lack of knowledge of these events; in particular, we have little idea as to how each of the endocrine or paracrine control systems interacts with the others under physiological conditions. This is primarily because our information is derived largely from *in vitro* findings and/or from studies of only a single hormonal process at a time. It is certain that our knowledge will improve dramatically over the next few years.

The importance of understanding the normal hormonal control of the testis is (1) that it may enable prediction of the effects of a given hormonal analog that has (paracrine) actions in the testis but also acts elsewhere in the body, and (2) that potentially it enables testicular function to be monitored via measurement of the various hormonal (endocrine and paracrine) markers, and such measurements may enable the primary testicular site of action of an administered compound to be identified. While admitting that, up to the present time, most of such markers have proved of relatively little use in this respect (but see Tables I and II), the exciting prospect for the next 5 years is that markers of paracrine function should become available, and these may permit monitoring of the minutiae of testicular function. At present, measurement of testicular weight and/or the serum levels of FSH provide the simplest and most accurate indication of seminiferous tubule dysfunction, but such measurements are insensitive and do not indicate specifically what has gone wrong, only that something has occurred that drastically impairs normal tubule function.

What the future may hold is perhaps best illustrated by a recent development. This showed that measurement in infertile men of the levels of the isoenzyme lactate dehydrogenase-C4 in semen in relation to sperm numbers provided an index of the efficiency of spermatogenesis in that individual, that is, of the degree of germ-cell degeneration occurring during spermatogenesis (Eliasson and Virji, 1985). Such measurements enabled subgroups of infertile men to be distinguished. The enzyme in question is a membrane component of postmeiotic germ cells and is of fundamental importance as it enables the germ cells to utilize the lactate and pyruvate, secreted by the Sertoli cell, as their energy source (see Chapter 2). But the real importance of such a marker is that it shows what insights into testicular function will become possible once suitable (paracrine) markers have been identified. Only then will we be able to pinpoint the primary testicular site of action of an administered compound or to identify the primary site of dysfunction in an infertile man. With such advances now clearly in prospect, the next 5–10 years promise to revolutionize our understanding of the hormonal control of the testis.

References

Aitken, R. J. (1983). The zona-free hamster egg penetration test. *In* "Male Infertility" (T. B. Hargreave, ed.), pp. 75–86. Springer-Verlag, Berlin.

Au, C. L., Robertson, D. M. and de Kretser D. M. (1983). *In vitro* bioassay of inhibin in testes of normal and cryptorchid rats. *Endocrinology* **112,** 239–244.

Bardin, C. W., Shaha, C., Mather, J., Salomon, Y., Margioris, A. N., Liotta, A. S., Gerendai, I., Chen, C-L, and Krieger, D. T. (1984). Identification and possible function of pro-opiomelanocortin derived peptides in the testes. *Ann. N.Y. Acad. Sci.* **438,** 346–364.

Bartlett, J. M. S., Kerr, J. B., and Sharpe, R. M. (1986). The effect of selective destruction and regeneration of rat Leydig cells on testicular morphology and the intratesticular distribution of testosterone and morphology of the seminiferous epithelium. *J. Androl.* **7,** 240–253.

Bellvé, A. R., and Feig, L. A. (1984). Cell proliferation in the mammalian testis: biology of the seminiferous growth factor. *Rec. Progr. Horm. Res.* **40,** 531–567.

Benahmed, M., Tabone, E., Reventos, J., and Saez, J. M. (1984). Role of Sertoli cells in Leydig cell function. *INSERM Colloq.* **123,** 363–385.

Bergh, A. (1983) Paracrine regulation of Leydig cells by the seminiferous tubules. *Int. J. Androl.* **6,** 57–65.

Boitani, C., Ritzen, E. M., and Parvinen, M. (1981). Inhibition of rat Sertoli cell aromatase by factor(s) secreted specifically at spermatogenic stages VII and VIII. *Mol. Cell. Endocrinol.* **23,** 11–22.

Catt, K. J., Harwood, J. P., Clayton, R. N., Davies, T. F., Chan, V., Katikineni, M., Nozu, K., and Dufau, M. L. (1980). Regulation of peptide hormone receptors and gonadal steroidogenesis. *Rec. Progr. Horm. Res.* **36,** 557–622.

de Kretser, D. M., and Kerr, J. B. (1983). The effect of testicular damage on Sertoli and Leydig cell function. *In* "Monographs on Endocrinology: The Pituitary and Testis" (D. M. de Kretser, H. G. Burger, and B. Hudson, eds.), pp. 133–154. Springer-Verlag, Berlin.

Eliasson, R., and Virji, N. (1985). LDH-C4 in human seminal plasma and its relationship to testicular function. II. Clinical aspects. *Int. J. Androl.* **8,** 201–214.

Fawcett, D. W., Neaves, W. B., and Flores, M. N. (1973). Comparative observations on intertubular lymphatics and the organization of the interstitial tissue of the mammalian testis. *Biol. Reprod.* **9,** 500–532.

Galdieri, M., Monaco, L., and Stefanini, M. (1984). Secretion of androgen binding protein by Sertoli cells is influenced by contact with germ cells. *J. Androl.* **5,** 409–415.

Hargreave, T. B., ed. (1983), "Male Infertility." Springer-Verlag, Berlin.

Janecki, A., Jakubowiak, A. and Lukaszyk, A. (1985). Stimulatory effect of Sertoli cell secretory products on testosterone secretion by purified Leydig cells in primary culture. *Mol. Cell. Endocrinol.* **42,** 235–243.

Jégou, B., Le Gac, F., Irby, D. C., and de Kretser, D. M. (1983a). Studies on seminiferous tubule fluid production in the adult rat: effect of hypophysectomy and treatment with FSH, LH and testosterone. *Int. J. Androl.* **6,** 249–260.

Jégou, B., Risbridger, G. P., and de Kretser, D. M. (1983b). Effects of experimental cryptorchidism on testicular function in adult rats. *J. Androl.* **4,** 88–94.

Jégou, B., Laws, A. O., and de Kretser, D. M. (1984). Changes in testicular function induced by short-term exposure of the rat testis to heat: further evidence for interaction of germ cells, Sertoli cells and Leydig cells. *Int. J. Androl.* **7,** 244–257.

Jutte, N. H. P. M., Jansen, R., Grootegoed, J. A., Rommerts, F. F. G., and van der Molen, H. J. (1983). FSH stimulation of the production of pyruvate and lactate by rat Sertoli cells may be involved in hormonal regulation of spermatogenesis. *J. Reprod. Fertil.* **68,** 219–226.

Kelch, R. P., Marshall, J. C., Sander, S., Hopwood, N. J., and Reame, N. E. (1983). Gonadotropin regulation during human puberty. *In* "Neuroendocrine Aspects of Reproduction in Primates" (R. L. Notman, ed.), pp. 229–256. Academic Press, New York.

Kerr, J. B., and Donachie, K. (1986). Regeneration of Leydig cells is stimulated by local testicular factors. *Cell Tissue Res.,* in press.

Kerr, J. B., and Sharpe, R. M. (1985). FSH induction of Leydig cell maturation. *Endocrinology* **116,** 2592–2604.

Kerr, J. B., and Sharpe, R. M. (1986). Effects and interactions of LH and LHRH on testicular morphology and function in hypophysectomized rats. *J. Reprod. Fertil.* **76,** 175–192.

Lincoln, G. A., and Short, R. V. (1980). Seasonal breeding: Nature's contraceptive. *Rec. Progr. Horm. Res.* **36,** 1–52.

Mason, A. J., Hayflick, J. S., Ling, N., Esch, F., Veno, N., Ying, S.-Y., Guillemin, R., Niall, H., and Seeburg, P. H. (1985). Complementary DNA sequences of ovarian follicular fluid inhibin show precursor structure and homology with transformin growth factor-*β*. *Nature (London)* **318,** 659–663.

Means, A. R., Fakunding, J. L., Huckins, C., Tindall, D. J., and Vitale, R. (1976). Follicle-stimulating hormone, the Sertoli cell and spermatogenesis. *Rec. Prog. Horm. Res.* **32,** 477–527.

Parvinen, M. (1982). Regulation of the seminiferous epithelium. *Endocr. Rev.* **3,** 404–417.

Parvinen, M., Vihko, K. M., and Toppari, J. (1987). Cell interactions during the seminiferous epithelial cycle. *Int. Rev. Cytol.,* in press.

Rea, M. A., Marshall, G. R., Weinbauer, G. F., and Neischlag, E. (1986). Testosterone maintains pituitary and serum FSH and spermatogenesis in gonadotrophin-releasing hormone antagonist-suppressed rats. *J. Endocrinol.* **108,** 101–107.

Rich, K. A., and de Kretser, D. M. (1983). Spermatogenesis and the Sertoli cell. *In* "Monographs on Endocrinology: The Pituitary and Testis" (D. M. de Kretser, H. G. Burger, and B. Hudson, eds.). pp. 84–105. Springer-Verlag, Berlin.

Russell, L. D. (1980). Sertoli-germ cell interrelations: A review. *Gamete Res.* **3,** 179–202.
Russell, L. D., Malone, J. P., and Karpas, S. L. (1981). Morphological pattern elicited by agents affecting spermatogenesis by disruption of its hormonal stimulation. *Tissue and Cell* **13,** 369–380.
Schanbacher, B. D. (1980). Androgen response of cryptorchid and intact rams to ovine LH. *J. Reprod. Fertil.* **59,** 151–154.
Setchell, B. P. (1978). "The Mammalian Testis." Paul Elek, London.
Sharpe, R. M. (1982). The hormonal regulation of the Leydig cell. *In* "Oxford Reviews in Reproductive Biology," Vol. 4 (C. A. Finn, ed.), pp. 241–317. Oxford Univ. Press, London and New York.
Sharpe, R. M. (1983). Local control of testicular function. *Q. J. Exp. Physiol.* **68,** 265–287.
Sharpe, R. M. (1984a). Intratesticular factors controlling testicular function. *Biol. Reprod.* **30,** 29–49.
Sharpe, R. M. (1984b). Intragonadal hormones (bibliography with review) *Bibliography of Reproduction* **44,** C1–C16.
Sharpe, R. M. (1986). Paracrine control of the testis. *Clin. Endocrinol. Metab.* **15,** 185–207.
Sharpe, R. M., and Bartlett, J. M. S. (1985). The intratesticular distribution of testosterone and the relationship to the levels of a peptide that stimulates testosterone production. *J. Reprod. Fertil.* **73,** 223–236.
Sharpe, R. M., and Cooper, I. (1983). Testicular interstitial fluid as a monitor for changes in the intratesticular environment in the rat. *J. Reprod. Fertil.* **69,** 125–135.
Sharpe, R. M., and Cooper, I. (1984). Intra-testicular secretion of a factor(s) with major stimulatory effects on Leydig cell testosterone secretion *in vitro. Mol. Cell. Endocrinol.* **37,** 159–168.
Sharpe, R. M., Cooper, I., and Doogan, D. G. (1984). Increase in Leydig cell responsiveness in the unilaterally cryptorchid rat testis and its relationship to the intratesticular levels of testosterone. *J. Endocrinol.* **102,** 319–327.
Sharpe, R. M., Kerr, J. B., Fraser, H. M., and Bartlett, J. M. S. (1986). Intratesticular factors and testosterone secretion: effect of treatments that alter the level of testosterone within the testis. *J. Androl.* **7,** 180–189.
Skinner, M. K., and Fritz, I. B. (1986). Identification of a non-mitogenic paracrine factor involved in mesenchymal–epithelial cell interactions between testicular peritubular cells and Sertoli cells. *Mol. Cell. Endocrinol.* **44,** 85–97.
Swerdloff, R. S., and Bhasin, S. (1984). Hormonal effects of GnRH agonist in the human male: An approach to male contraception using combined androgen and GnRH agonist treatment. *In* "LHRH and Its Analogues" (F. Labrie, A. Belanger, and A. Dupont, eds.), pp. 287–301. Elsevier, Amsterdam.
Swerdloff, R. S., and Heber, D. (1981). Endocrine control of testicular function from birth to puberty. *In* "Comprehensive Endocrinology: The Testis" (H. Burger and D. M. de Kretser, eds.), pp. 107–126. Raven, New York.
Ultee-van Gessel, A. M., Leemborg, F. G., de Jong, F. H., and van der Molen, H. J. (1986). *In-vitro* secretion of inhibin-like activity by Sertoli cells from normal and prenatally irradiated immature rats. *J. Endocrinol.* **109,** 411–418
Vihko, K. K., Suominen, J. J. O., and Parvinen, M. (1984). Cellular regulation of plasminogen activator secretion during spermatogenesis. *Biol. Reprod.* **31,** 383–389.
Waites, G. M. H., and Gladwell, R. T. (1982). Physiological significance of fluid secretion in the testis and blood–testis barrier. *Physiol. Rev.* **62,** 624–671.
Wang, J., Galil, K. A. A., and Setchell, B. P. (1983). Changes in testicular blood flow and testosterone production during aspermatogenesis after irradiation. *J. Endocrinol.* **98,** 35–46.

Wathes, D. C. (1984). Possible actions of gonadal oxytocin and vasopressin. *J. Reprod. Fertil.* **71,** 315–345.

Worley, R. T. S., Nicholson, H. D., and Pickering, B. T. (1985). Testicular oxytocin: An initiator of seminiferous tubule movement? *INSERM Colloq.* **123,** 205–212.

Wright, W. W., Parvinen, M. M., Musto, N. A., Gunsalus, G. L., Philips, D. M., Mather, J. P., and Bardin, C. W. (1983). Identification of stage-specific proteins synthesized by rat seminiferous tubules. *Biol. Reprod.* **29,** 257–270.

Wu, F. C. W. (1985). Male hypogonadism—current concepts and trends. *Clin. Obstet. Gynec.* **12,** 531–555.

Yee, J. B., and Hutson, J. C. (1985). Effects of testicular macrophage—conditioned medium on Leydig cells in culture. *Endocrinology* **116,** 2682–2684.

Chapter 5

Male Neuroendocrinology and Endocrine Evaluation of Reproductive Disorders

ANDRES NEGRO-VILAR AND MARCELO M. VALENCA

Reproductive Neuroendocrinology Section, L.R.D.T.
National Institute of Environmental Health Sciences
National Institutes of Health
Research Triangle Park, North Carolina 27709

I. Introduction

The two major functions of the testis, gametogenic and endocrine, while regulated by endocrine input provided by the hypothalamic–pituitary axis, are in effect under separate controlling mechanism(s) that regulate the activity of each component. Although the gametogenic and endocrine activities of the testis are not independent under physiological conditions, experimental or clinical disorders can indeed affect normal function of one with little or no effect on the other. As in many other endocrine-regulated systems, the function of the male gonad depends on a highly complex series of feedback mechanisms taking place at both central and peripheral sites. The brain participates in this regulation through the production by the hypothalamus of luteinizing hormone–releasing hormone (LHRH), a peptidergic hormone produced by neurons located in the hypothalamic–preoptic regions. This peptide, when released into the hypophyseal–portal circulation,

PHYSIOLOGY AND TOXICOLOGY
OF MALE REPRODUCTION

reaches the anterior pituitary, binds to specific membrane receptors in the gonadotrophs, and enhances the release of both gonadotropins, luteinizing hormone (LH) and follicle-stimulating hormone (FSH), from the anterior pituitary into the general circulation. The LH binds to specific receptors in the Leydig cells of the testis, and this leads to stimulated androgen production. The FSH acts on the seminiferous tubular epithelium to stimulate spermatogenesis (in conjunction with androgens). The endocrine feedback axis is completed by the action of circulating testicular hormones on brain and pituitary sites to modulate and/or modify the output of LHRH/gonadotropins from the hypothalamic–pituitary axis. Dysfunction at any level of the system can result in either selective or generalized malfunction of the gonadal endocrine and/or gametogenic functions. This chapter will deal primarily with some of the intimate mechanisms operating at both the hypothalamic and the pituitary level to regulate gonadotropin release and, ultimately, gonadal function. We will describe initially the operating characteristics of both levels of control, followed by an analysis of the potential targets for reproductive dysfunction as well as guidelines for the endocrine exploration of the infertile male.

II. Brain Control of Reproductive Functions: The Role of Luteinizing Hormone–Releasing Hormone

Since the discovery of LHRH (GnRH), we have witnessed great advances in our knowledge of the anatomy, physiology, pharmacology, and clinical applications of the decapeptide. These advances were greatly stimulated by the isolation and sequencing of the decapeptide by Schally *et al.* (1971). The development of microdissection techniques (Arimura, 1976; Setalo *et al.*, 1976; Palkovits, 1978; Witkin *et al.*, 1982; Anthony *et al.*, 1984) greatly contributed to the mapping of the distribution and localization of LHRH within the different brain regions. Another important contribution to the study of LHRH function resulted from the development of different peptidic analogs of LHRH, possessing either agonist or antagonistic properties. The availability of LHRH agonists resistant to enzymatic degradation allowed different groups to provide a detailed description of the pituitary and gonadal LHRH receptor under different physiological and pathological conditions (Clayton and Catt, 1980; Marshall *et al.*, 1981). Synthetic LHRH analogs have been found useful in the treatment of selected cases of infertility (Schally *et al.*, 1978; Crowley and McArthur, 1980), and under certain regimens of administration LHRH analogs have been found to have paradoxical antifertility effects (Sandow, 1986; Baja and Madan, 1983). This property of the agonists may prove useful for these peptides as contraceptive agents. Moreover, recent studies have described the presence of an LHRH-like peptide in the testis (Sharpe *et al.*, 1982; Bhasin *et al.*, 1983), and direct gonadal effects of LHRH and its analogs have also been reported (Sharpe *et al.*, 1982).

It is evident from these recent advances that LHRH plays a major modulatory role in the regulation of gonadal function, acting at different levels within the hypothalamic–pituitary–gonadal axis. A clear understanding of the regulatory mechanisms involved in modulation of release and action of LHRH is of great significance for our knowledge of the physiology and pharmacology of the decapeptide, for its potential uses and applications in fertility and infertility problems, and for the toxicological evaluation of potential hazardous chemicals and environmental agents.

A. Anatomical Arrangement of the Hypothalamic–Pituitary Axis: Functional Aspects

1. Luteinizing Hormone–Releasing Hormone

Several immunocytochemical studies have demonstrated that LHRH cell bodies are widely distributed in most species within rostral preoptic and hypothalamic regions, with a small number of neuronal cell bodies located in other brain regions (Setalo *et al.*, 1976; King *et al.*, 1982; Witkin *et al.*, 1982; King and Anthony, 1983; Sar *et al.*, 1987). The decapeptide is produced within the neuronal cell body (King *et al.*, 1982; Sar *et al.*, 1987) and is transported axonally to the nerve terminals, many of which are located in the region of the median eminence. Other terminal fields have also been described, and LHRH neurons have been reported to make synaptic contacts with other neurons (Marshall and Goldsmith, 1980; Witkin *et al.*, 1982; Liposits *et al.*, 1984), as well as with LHRH cells, which may suggest a possible autoregulatory function. This seems to be supported by recent findings from our laboratory supporting direct central effects of LHRH analogs on endogenous LHRH neuronal systems. This observation seems to substantiate the idea that LHRH neurons may have, under certain conditions, an autofeedback regulatory mechanism. Within the region of the median eminence, LHRH fibers are distributed largely in the external zone, in close contact with the portal capillaries, although in certain species such as human and monkey a considerable number of fibers have been found in the internal zone of the median eminence. In most species, including humans, a good number of LHRH fibers project to the neural lobe and terminate within the neural lobe in close proximity to the anterior pituitary (Anthony *et al.*,

1984). The exact role of these projections remains unknown, but they could represent an alternative pathway of delivery of the decapeptide to the adenohypophysis.

2. Distribution of the Luteinizing Hormone–Releasing Hormone Prohormone

Recently, the structure of the cDNAs encoding human and rat LHRH-prohormone were determined and the corresponding peptide sequences deduced (Adelman *et al.*, 1986). In addition to LHRH, both cDNAs also encode a 56-amino acid peptide sequence, designated as gonadotropin-releasing hormone associated peptide (GAP) (Adelman *et al.*, 1986; Nikolics *et al.*, 1985), with approximately 70% homology existing between the human and rat GAP sequences. Antisera generated against a fragment of the human GAP sequence (hGAP 25-53; proLHRH 38-66) (Culler and Negro-Vilar, 1986a) contain antibody populations that are specific for antigenic sites within the midportion of the hGAP sequence and that can recognize the entire hGAP 1–56 sequence, as well as GAP-like immunoreactivity in rat hypothalamic extracts (Culler and Negro-Vilar, 1986a). These antisera were used to study the localization of GAP-like immunoreactivity in the rat brain by the avidin–biotin complex technique (ABC). Immunostaining of neuronal perikarya, fibers, and terminals was demonstrated with GAP antiserum under conditions of tissue preparation that make immunostaining with LHRH antisera difficult or undetectable. The GAP- immunoreactive perikarya were observed in sections of perfused or nonperfused brains without colchicine pretreatment (Sar *et al.*, 1987). Using a double immunoperoxidase staining method, both GAP and LHRH immunoreactivities were shown to coexist in the same neurons. The common distribution of LHRH and GAP immunoreactivity in the rat brain is strongly supportive of the GAP representing the non-LHRH portion of the LHRH precursor. The use of GAP antisera that can distinguish between the LHRH and the remaining portion of its prohormone represents a valuable tool for studies of the LHRH-prohormone processing and distribution (Sar *et al.*, 1987). Although not yet confirmed, the human GAP sequence has been reported to possess both gonadotropin-releasing activity and prolactin-inhibiting activity (Nikolics *et al.*, 1985). Additionally, at least one peptide sequence contained within the human GAP sequence (GAP 1–13 or pro-LHRH 14–26) has been reported to stimulate gonadotropin release (Millar *et al.*, 1986). Regardless of whether the non-LHRH portion of the LHRH prohormone contains biological activity, this sequence can serve as a valuable marker for studies of LHRH prohormone synthesis, processing, and degradation.

B. Hypothalamic Control of Gonadotropin Secretion

1. Monoaminergic Systems Regulating Luteinizing Hormone–Releasing Hormone and Gonadotropin Release

One of the central questions in neuroendocrinology concerns the role played by the different neurotransmitter systems in the regulation of gonadotropin secretion. A better understanding of the complex interactions between the regulatory neurons, with a diversity of chemical neurotransmitters, and the hypophysiotrophic (peptidergic) neurons, which act as neurotransducers to transform an electrochemical impulse into a hormonal signal, may provide valuable clues concerning the mechanism of neuronal communication. The neural mechanisms that control the release of hypophysiotrophic peptides from peptidergic neurons are not yet clearly understood. The hypothalamic peptidergic neurons receive numerous synaptic inputs both from conventional and from putative neurotransmitters. Presynaptic regulation of neurotransmitter release has been described for the adrenergic system in both peripheral and central neurons. From ultrastructural and neurochemical studies, it seems logical to infer that similar neurochemical regulatory mechanisms may be operative at the presynaptic peptidergic nerve terminals in the hypothalamus. Indeed, the complexity of the system is overwhelming. Redundancy is a common feature in the regulation of reproductive functions, with multiple and generally diffuse pathways involving different types of neurotransmitters being called on to participate in the modulation of the events. Elimination of one of those pathways, or of one of the neurotransmitter systems involved, sometimes results in only partial or temporary interruption or delay of a given function. It is obvious, then, that to define the role of a given neurotransmitter or a neuronal pathway in a specific regulatory mechanism, it is necessary to evaluate carefully what other pathways or systems are involved and what effects the interference with one of those systems may have on the function of the others and, finally, on the overall response being studied. Neuroanatomical and neurochemical mapping of the distribution of monoaminergic and peptidergic systems in different brain regions have provided valuable information concerning specific neurotransmitter systems and their involvement in the regulation of well-defined physiological events.

The hypothalamic peptidergic neurons are in synaptic contact with a host of putative synaptic transmitters (Fig. 1). Noradrenergic cell bodies located in the medulla oblongata and pons give rise to the two major ascending noradrenergic pathways, one of which, the ventral noradrenergic tract, innervates the whole hypothalamus, the internal layer of the median eminence (ME), and the retrochiasmatic and preoptic areas (Ungerstedt, 1971). An-

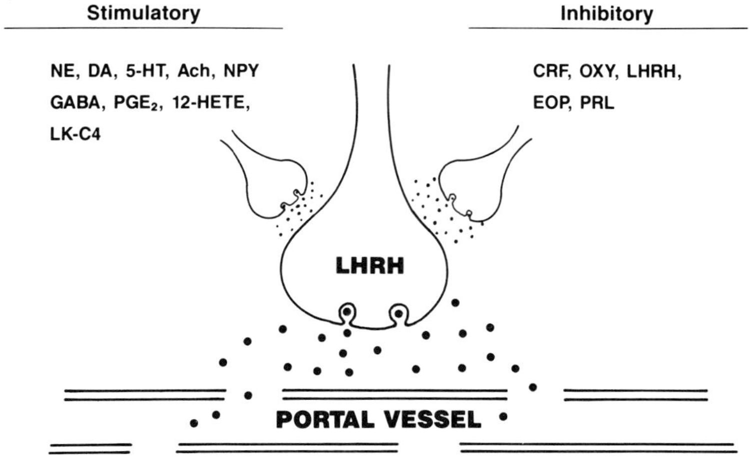

Fig. 1. This diagram shows a number of neurotransmitters or neuromodulators involved in the neuroendocrine regulatory events controlling the secretion of LHRH from median eminence nerve terminals into the hypophyseal portal blood. Although many endogenous substances influence the secretion of gonadotropins, the brain messengers shown here are those that most clearly stimulate (left axon terminal) or inhibit (right axon) LHRH release *in vitro* or *in vivo* into the portal vessels. Some agents such as DA or GABA have been reported to cause both stimulatory and inhibitory activity (see text). NE, norepinephrine; DA, dopamine; 5-HT, serotonin; Ach, acetylcholine; NPY, neuropeptide Y; GABA, γ-aminobutyric acid; PGE_2, prostaglandin E_2; 12-HETE, 12-hydroxyeicosatetraenoic acid; LK-C4, leukotriene C4; CRF, corticotropin-releasing factor; OXY, oxytocin; LHRH, luteinizing hormone-releasing hormone; EOP, endogenous opioid peptides; PRL, prolactin. For further explanation of these agents see text.

other noradrenergic tract originates in the locus coeruleus and projects rostrally as the dorsal noradrenergic bundle to innervate the dorsomedial, paraventricular, and periventricular hypothalamic nuclei.

The tuberoinfundibular dopaminergic system has its cell bodies located in the arcuate nuclei, with the majority of its fibers terminating in the ME and the neural lobe. A secondary group of dopamine (DA) cell bodies has been described (Bjorklund *et al.*, 1975) in the posterior hypothalamus and zona incerta, with axons projecting to the anterior hypothalamus and suprachiasmatic region. The function of this tract is unknown, but because it innervates hypothalamic regions involved in the preovulatory discharge of gonadotropins and prolactin, it may play an important role in mediating these changes. Epinephrine-containing neurons have been located in the brainstem with axons that project rostrally to terminate in the hypothalamus in the region of the arcuate, ventromedial, and dorsomedial nuclei (Hokfelt *et al.*, 1974).

Serotoninergic neurons in the medial raphe nuclei project to the suprachiasmatic, anterior hypothalamic, and median eminence regions (Saavedra *et al.*, 1974). Assays for choline acetyltransferase indicate the widespread distribution of cholinergic fibers within the hypothalamus as well. There is also an abundance of histamine that seems to be located in synaptosome-like structures; this amine is particularly concentrated in the median eminence region, where it may also serve as a synaptic transmitter (Brownstein *et al.*, 1974).

An increasing number of neurophysiological and, more recently, neuroanatomical and biochemical studies have shown that certain amino acids may function as transmitter substances in the central nervous system (CNS). Premier candidates for such a role are γ-aminobutyric acid (GABA) and glycine. Other amino acids, however, such as proline, taurine, alanine, glutamic acid, and aspartic acid, have also been suggested as possible neurotransmitters. Several of these amino acids were found to be involved, either directly or indirectly, in the release of different neural peptides.

The central noradrenergic system plays a major role in the regulation of the LHRH release, and many reports have substantiated this concept. Blockade of norepinephrine (NE) synthesis prevents the preovulatory proestrous surge of LH (Kalra and McCann, 1974), the LH release induced by progesterone in estrogen-primed ovariectomized rats (Kalra and Kalra, 1984), and the early rise in gonadotropins observed after orchidectomy (Ojeda and McCann, 1974). Blockade of α-adrenergic receptors (Kalra and Kalra, 1984) yielded similar results. Intraventricular injection of NE reportedly elicits LH release in the ovariectomized or in ovariectomized estrogen/progesterone-treated (OEP) animals (Vijayan and McCann, 1974; Kreig and Sawyer, 1976). An acute increase in NE content in the ME of orchidectomized males preceded the rise in gonadotropins by several hours. This change was reversed by acute testosterone treatment, with doses that effectively inhibited the postcastration rise of gonadotropins (Chiocchio *et al.*, 1976). An increase in NE content and turnover in the ME, at the time of the critical period that precedes the preovulatory discharge of LH, has been documented (Negro-Vilar *et al.*, 1977a; Lofstrom, 1977).

In ovariectomized rats interference with the noradrenergic system abolished the circhoral LH pulses and decreased LH levels (Gnodde and Schuiling, 1976; Drouva and Gallo, 1976). Also destruction of the ventral noradrenergic pathway reduced or abolished the LH surge in proestrous rats (Martinovic and McCann, 1977). In the rat, bilateral deafferentation of the hypothalamus drastically reduced NE content in the ME, which indicates that noradrenergic axons innervating the ME enter the hypothalamus from its lateral aspects. On the other hand, similar treatment did not prevent the postcastration rise in LH (Gallardo *et al.*, 1978). Another reported study provided further evidence in support of the prominent role played by the noradrenergic system on the postcastration rise in gonadotropins, particularly in LH (Advis *et al.*, 1980). These results clearly demonstrated that the negative feedback actions of estradiol are mediated, at least in part, through central inhibition of the noradrenergic system in specific brain areas. Also demonstrated was the involvement of the suprachiasmatic–medial preoptic region in the negative feedback action of estradiol, a surprising discovery since this area had generally been implicated in the LH release induced by the positive feedback effect of steroids. More conclusive evidence for a stimulatory role of NE on LHRH release from MBH or ME nerve terminals had previously been demonstrated by incubating these hypothalamic fragments in presence of NE and measuring LHRH release by RIA (Negro-Vilar *et al.*, 1979). That stimulatory effect of NE on LHRH release *in vitro* is mediated through an α-adrenergic receptor, since blockade of this receptor by phentolamine nullified such effect.

A stimulatory role of DA on LH release is supported by the observation that increased LH levels are detected in estrogen–progesterone-treated rats after intraventricular (Vijayan and McCann, 1978a) or systemic (Vijayan and McCann, 1978b) injection of DA or the DA-receptor agonist, apomorphine. Similarly, the proestrous LH surge was advanced by administration of a DA-receptor agonist (Clemens *et al.*, 1977). Other reports suggest an inhibitory role of DA on LH secretion. Pharmacologic blockade of DA receptors in ovariectomized rats either inhibited (Gnodde and Schuiling, 1976) or had no effect (Drouva and Gallo, 1976) on LH levels. On the other hand, apomorphine was reported to suppress pulsatile LH secretion in castrated rats (Gnodde and Schuiling, 1976; Drouva and Gallo, 1976). The DA turnover in the ME was found to decrease at the time of the proestrous surge (Löfström, 1977). Under certain conditions, DA can release LHRH from synaptosomal preparations of sheep or rat hypothalamus (Bennett *et al.*, 1975), as well as from rat MBH fragments incubated *in vitro* (Rotsztejn *et al.*, 1976). In our hands, DA was found to stimulate LHRH release from ME fragments (Negro-Vilar *et al.*, 1979) in a dose-related manner, an effect blocked by the DA-receptor blocker, pimozide.

The cholinergic system may also play an important role in the control of LHRH release. Intraventricular administration of acetylcholine can increase LH release in ovariectomized animals (Fiorindo *et al.*, 1974; Vijayan and McCann, 1980). This effect of acetylcholine is probably mediated by muscarinic receptors, since atropine can block the effects of the amine and also lowers basal LH levels in the castrated male. An interaction between the cholinergic and the dopaminergic system to regulate LH secretion has also been suggested (Vijayan and McCann, 1980).

2. Role of GABA in the Regulation of Luteinizing Hormone–Releasing Hormone and Luteinizing Hormone Release

Evidence accumulated during the last decade suggests that γ-aminobutyric acid (GABA) may play a fundamental role at the hypothalamic–pituitary level to control pituitary hormone secretion (Muller *et al.*, 1983; Masotto and Negro-Vilar, 1986).

In terms of gonadotropin secretion, the role of GABA and its analogs has not yet been fully clarified, with some studies supporting a stimulatory effect of GABA on gonadotropin release, and others favoring an inhibitory action. An intrinsic hypothalamic GABAergic pathway has been described, with cell bodies in the arcuate nucleus and terminals in the ME and in rostral regions containing the LHRH cell bodies (Muller *et al.*, 1983; Leranth *et al.*, 1985). The primary effect of GABA *in vivo* seems to be inhibitory, and this effect may involve interactions with noradrenergic neurons (Wuttke *et al.*, 1980). A role of steroids in mediating inhibitory or excitatory responses to GABA has been suggested (Masotto and Negro-Vilar, 1986).

3. Role of Endogenous Opiate Peptides on the Regulation of Luteinizing Hormone–Releasing Hormone and Luteinizing Hormone Release

The endogenous opiate neuronal systems are important modulators of the different hypothalamic peptidergic systems that regulate pituitary function (Meites *et al.*, 1983). In the reproductive area, endogenous opiates have been found to provide a tonic inhibitory tone over gonadotropin secretion (Meites *et al.*, 1983; Kalra and Kalra, 1984). Blockade of opiate receptors in both animals and humans results in a rapid rise in LH secretion. Several studies suggest that this effect is centrally mediated and that catecholamines may be involved in this response, since inhibition of norepinephrine and epinephrine synthesis nullifies the effect of the opiate receptor blocker naloxone (NAL) (Kalra and Kalra, 1984).

Using an *in vitro* system that we have characterized for the evaluation of LHRH release, we determined recently that NAL can promptly increase LHRH release from hypothalamic fragments consisting of the arcuate–median eminence (ARC-ME) region (Negro-Vilar and Valenca, 1985). This fragment contains the intact cell bodies of β-endorphin and dopamine neurons located in the arcuate nucleus, as well as all the normal anatomical connections innervating LHRH fibers and terminals within the basal hypothalamic region. Most LHRH cell bodies, on the other hand, are not included in the dissected fragment. The increase in LHRH after NAL indicates that removal of opiatergic inhibition within the ARC-ME region can account for the increased release of LHRH (and subsequently of LH) seen *in vivo.* Because of the anatomical characteristic of the incubated fragments, the opiatergic inhibition must be exerted, directly or through an interneuron, on the LHRH axons and terminals. Further studies indicated that release of prostaglandin-E2 (PGE2), a well-known secretagogue for LHRH, is also increased after NAL is added *in vitro* to ARC-ME fragments. Addition of an α-receptor blocker, phentolamine, prior to NAL blocks the increase in both PGE2 and LHRH (Negro-Vilar *et al.*, 1985). Previous studies have indicated that norepinephrine increases LHRH release by interacting with an α-receptor located in nerve terminals in the median eminence (Ojeda *et al.*, 1982). As a result of this interaction, PGE2 is released, and this in turn results in release of LHRH by mobilization of intracellular Ca^{2+}. The present results indicate that opiate (possibly β-endorphin) neurons exert a tonic-inhibitory tone on catecholaminergic elements within the arcuate–median eminence region, and this results in a decreased release of PGE2, a key intracellular mediator of LHRH release (Negro-Vilar *et al.*, 1985).

C. *Feedback Mechanisms Modulating Luteinizing Hormone–Releasing Hormone and Luteinizing Hormone Release*

1. Long-Loop Feedback: Gonadal Steroids and Peptides

Feedback interactions take place at the level of the anterior pituitary (in addition to the hypothalamic level) to either inhibit or enhance gonadotropin release (Fig. 2). An inhibitory feedback of the testicular hormones is more frequently observed in males. Castration results in elevation of both LH and FSH levels, whereas substitutive therapy with either androgens or estradiol can restore gonadotropin levels to normal. From animal studies it becomes clear that these effects are exerted at both the hypothalamic and pituitary levels. Either aromatizable (testosterone) or nonaromatizable (dihydrotestosterone) androgens can suppress gonadotropin release, suggesting that the effects of testosterone are not simply due to its conversion to estrogens. Moreover, estrogens act more rapidly than testosterone to decrease pituitary responsiveness to LHRH, suggesting a different mechanism of action.

Responsiveness of the pituitary gland to LHRH administration can be affected by different factors. Steroids can both diminish and enhance pituitary sensitivity to the decapeptide. Similarly, certain androgens can differentially modulate pituitary responsiveness to LHRH, so that selective effects in either LH or FSH can be obtained (Negro-Vilar and DePaolo, 1981; Santen, 1977). Positive feedback effects can be observed after estradiol treatment. Although this effect is more clearly seen in women, some effects can be observed in adult males, and more clearly in patients with hypergonadotrophic hypogonadism.

As indicated, the inhibitory effect of gonadal steroids on pituitary response to LHRH is well established (Negro-Vilar, 1973; Davidson, 1969). The effect of testosterone is always clearly inhibitory (Davidson *et al.*, 1976), whereas that of estrogen and progesterone may vary according to the time elapsing between the administration of the hormone and/or the prevailing endocrine condition of the animal (Fink, 1979). It seems clear, therefore, that the negative feedback effect of gonadal steroids is exerted at both hypothalamic and pituitary levels and that the so-called tonic release of pituitary

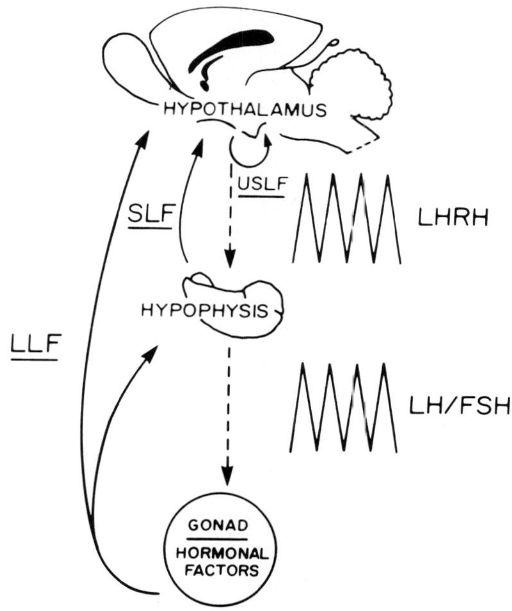

Fig. 2. This diagram demonstrates the feedback control mechanisms modulating the hypothalamic–pituitary–testicular axis. The neurohormone LHRH is released in a pulsatile pattern from median eminence nerve endings, which results in the episodic or pulsatile secretion of LH. The pulsatile secretion of FSH appear to be under separate hypothalamic control. The actions of the gonadotropins on the testes promote the production of sex steroids and other gonadal factors, which, in turn, through a long loop feedback (LLF) mechanism, influence brain LHRH secretion. In addition, LHRH may negatively regulate its own release by an ultrashort-loop feedback (USLF) mechanism. On the other hand, the gonadotropins may also influence LHRH secretion through a short-loop feedback (SLF) mechanism.

gonadotropins is the result of the relative preponderance of a controlling signal (LHRH) and the modulatory influence of an inhibitory gonadal input (steroids).

Concerning the stimulatory feedback effect of steroids, there is clear evidence that the preovulatory surge of gonadotropin is triggered by an increase in circulating estradiol levels. Characterization of this stimulatory effect of estrogen in the monkey indicates that blood levels of the steroid must reach a certain threshold and remain elevated for no less than 24 h (Karsch *et al.*, 1973) in order to trigger an LH surge. In the rat, progesterone can facilitate the stimulatory effect of estrogen on gonadotropin release (Brown-Grant, 1976), but the marked increase in progesterone that follows the preovulatory LH surge, rather than triggering another LH increase, prevents subsequent LH surges within the same cycle (Freeman *et al.*, 1976). In the absence of such an "extinction" signal provided by progesterone, repeated LH surges occur.

In contrast with the steroid negative feedback mechanism, the positive feedback occurs only in females, is primarily induced by estrogen, and cannot be generated by testosterone. Both feedbacks, however, have in common a pituitary and a neural substrate for their effects.

These pituitary and hypothalamic feedback effects of the steroids constitute what is known as the long-feedback loop, reflecting the gonad to brain (and pituitary) axis. In addition to the effects of the steroids, another important source of feedback signals is represented by the gonadal peptides. Among those, the role of inhibin has been clearly defined as an important gonadal peptide specifically modulating the secretion of FSH.

McCullagh (1932) described and assigned the name inhibin to a nonsteroidal, water-soluble principle of testicular origin, which could prevent the appearance of castration cells in the pituitary gland of gonadectomized animals. This term has been adopted for any nonsteroidal gonadal factor that exerts an inhibitory action on the release of FSH. Others have defined inhibin more specifically as a peptidic factor of gonadal origin that specifically or selectively suppresses FSH secretion. Other terms that have been used to define this (or similar) substance(s) are folliculostatin, follitropin-suppressing principle, and Sertoli cell factor. The fact that several terms have been used in the past to describe this substance reflects the different sources of tissues and animals from which inhibin principles have been purified. Preparations from human and bull seminal plasma, ram rete testis fluid, bovine and porcine follicular fluid, and rat Sertoli or granulosa cells have all been found to have FSH-suppressing activity (Franchimont *et al.*, 1979; Negro-Vilar and Lumpkin, 1983).

The different preparations used have been shown to suppress basal or stimulated FSH release in a variety of *in vivo* or *in vitro* models. These actions of inhibin are, in most cases, selective for FSH. Only in certain paradigms was LH release also affected. From the many experiments performed over the last decade, a general consensus has emerged indicating that one of the primary sites of action of inhibin to suppress FSH release is at the pituitary level. Intraperitoneal injection of bovine seminal plasma into normal and castrated male rats lowered basal FSH levels in plasma but had no effect on LH values (Franchimont *et al.*, 1979). Pretreatment of male rats with inhibin decreased FSH release in response to an LHRH injection, but failed to alter the release of LH stimulated by LHRH administration (Franchimont *et al.*, 1979). Using the proestrous female rat in which the preovulatory LH and FSH surges were blocked with phenobarbital, Wise *et al.*, (1979) were able to suppress LHRH-induced FSH secretion by porcine follicular fluid, a source of inhibin, without noticeably altering the LH response. That these different inhibin preparations were able to alter the pituitary responsiveness of FSH to LHRH *in vitro* and *in vivo* establishes a pituitary site of action for this gonadal substance. Inhibin has been shown to have also a central site of action to inhibit selectively FSH release. Intraventricular administration of purified inhibin preparations obtained from ram rete testis fluid produced a very marked inhibition of FSH release, without affecting LH levels, in orchidectomized male rats. The effect was shown to be mediated centrally, since pituitary responsiveness to LHRH was not impaired (Lumpkin *et al.*, 1981). Whether the central action of inhibin is physiologically relevant remains to be determined.

Recently, the structure of inhibin has been elucidated. Inhibin was isolated and purified from the gonadal fluids of several species, and characterized as a heterodimeric protein consisting of α and β polypeptide chains associated by disulphide bonds (Mason *et al.*, 1985). Two forms of inhibin (A and B) were characterized, each containing an identical α subunit and a distinct but related β subunit. More recently, another heterodimer formed by the β subunits from the two forms of inhibin has been described. This molecule, named activin or FSH-releasing protein (FRP), has an opposite effect to that of inhibin—that is, it stimulates FSH release (Ling *et al.*, 1986); Vale *et al.*, 1986). Activin has a good degree of homology with transforming growth factor β (TGF-β), and it is now clear that other proteins showing homology, such as Mullerian-inhibin substance, may be products from a gene family with important roles in the regulation of cell differentiation, cell growth, and control of reproductive functions.

With gonadal peptide molecules capable of inhibiting and stimulating gonadotropin secretion, the concept of long-feedback loops has to be expanded to include these factors, side by side with the well-known effects of the gonadal steroids.

2. Short-Loop Feedback: Pituitary Hormones Affecting Luteinizing Hormone–Releasing Hormone Release

Although possible direct effects of gonadotropic hormones on LHRH secretion have been described, no clear evidence of a physiological role of LH and FSH in central regulation of LHRH release is yet available. On the other hand, other pituitary hormones, most notably prolactin, have been shown to profoundly affect gonadotropin secretion and reproductive functions (Negro-Vilar, 1980; Bartke, 1986) (Fig. 2).

Although prolactin is under dual hypothalamic control, an inhibitory tone normally prevails. Both a prolactin-inhibiting factor (PIF) and a prolactin-releasing factor (PRF) have been described as putative hypothalamic hormones involved in the physiological regulation of prolactin. Neither of these factors has been isolated, and their exact chemical nature is unknown. However, dopamine, a catecholaminergic neurotransmitter present in high amounts in the hypothalamus, has been shown to have clear prolactin-inhibitory activity, and is thought by many to be the PIF. A recent report (Nikoliks *et al.*, 1985) ascribing PIF activity to the non-LHRH portion of the LHRH prohormone (GAP) has remained unconfirmed by other laboratories. Regardless of the true chemical nature of PIF, it is clear that the hypothalamic inhibitory tone over the release of prolactin is mediated primarily through or by dopamine. This firmly established concept has important clinical and therapeutical implications. A variety of centrally acting drugs widely used in the clinic is known to interfere with central dopaminergic metabolism when given in large doses or for prolonged periods of time, thereby interfering with the normal dopaminergic regulation of prolactin. Abnormally high prolactin levels have been reported in patients after treatment with phenothiazines (chlorpromazine, thioridazine, etc.), reserpine, α-methyldopa, and a number of other neuroleptic drugs. In turn, dopaminergic agonists (like 2-bromoergocryptine) or L-DOPA (a dopaminergic precursor clinically used for the treatment of Parkinson's disease) are potent inhibitors of prolactin release. It is not surprising, then, that chronic treatment with some of these drugs may lead to disorders in reproductive functions in both men and women.

Like other pituitary hormones, prolactin in humans is secreted episodically. Circadian variations in prolactin secretion have been found to be sleep-entrained, with the highest circulating concentrations of prolactin occurring during the nocturnal sleep period, particularly in the early morning hours. These diurnal prolactin fluctuations have been found to be present in prepubertal and pubertal boys as well as in adult men. In adult males, hyperprolactinemia has been recognized with increasing frequency in recent years, with one-third of the patients showing galactorrhea, and some loss of libido and lack of erections. Some of these patients present smaller prostates and a decreased volume of their seminal ejaculate. Treatment with bromoergocryptine can restore potency and libido and suppress the galactorrhea concomitantly with the lowering of serum prolactin levels (Thorner *et al.*, 1977). Some reports indicate that oligozoospermia and infertility can occur in patients with high serum prolactin levels. In aging male rats, a progressive increase in prolactin levels has been reported combined with a gradual decline in fertility, as well as with decreased secretion of LH, FSH, and testosterone. This progressive increase in prolactin during aging may be related to a decreased dopaminergic and increased serotoninergic metabolism in the hypothalamus. There seems to be good agreement in that a primary target for the effect of high prolactin levels is the LHRH–neuronal system, and that decreased LHRH release, coupled with an increased sensitivity to negative feedback by gonadal steroids, is responsible for the suppressed gonadotropin secretion and for the resulting hypogonadism (Bartke, 1986). In support of this concept is the recent demonstration of reduction in frequency of pulsatile LH episodes in hyperprolactinemic men. Whether the effects of prolactin can be exerted in the LHRH neurons directly is still not clear. However, it has been well established that many of the central effects of prolactin are mediated by its interaction with dopaminergic and perhaps also with opiatergic neurons. There is also evidence for peripheral effects of prolactin, at both the pituitary and testicular level. Prolactin has been reported to decreased pituitary sensitivity to LHRH, and it has also been shown to affect the capacity of the testis to produce testosterone and also to affect the conversion of testosterone to other metabolites. The mechanisms responsible for the decreased libido and potency are yet to be clearly defined.

Corticotropin (ACTH) and β-endorphin are two additional pituitary hormones that can negatively affect gonadotropin secretion. In situations of prolonged or intense stress, the highly elevated levels of these two hormones can possibly affect gonadal function. As discussed above, β-endorphin from a central origin is an important neural factor with a potential tonic inhibitory role on LHRH release. Corticotropin-releasing factor (CRF) has been shown to inhibit LHRH and LH release, by mechanisms mediated at a central site and not involving ACTH. Again, chronic stress may be a situation where this mechanism could be operative to block gonadotropin release and interfere with normal reproductive functions.

3. Ultrashort-Loop Feedback

The concept of ultrashort-loop feedback was first introduced by Martini, to indicate the possibility that a peptide secreted from a neuron may modify the activity of the neuronal system producing and secreting a hypophysiotrophic hormone. Recently, evidence has been obtained to suggest that this ultrashort-feedback system may be operative within the LHRH neuronal system. LHRH agonists, given in a regimen known to suppress gonadal function, have been shown to inhibit endogenous LHRH secretion into the hypophyseal portal circulation, by mechanisms involving a central (hypothalamic) site of action (Valenca *et al.*, 1986). The LHRH antagonists that can effectively block LHRH receptors produced a significant increase in basal LHRH release *in vitro*. These observations, coupled with the anatomical evidence suggesting LHRH synaptic contacts with LHRH neurons, strongly suggest a functional role for LHRH neurons to modulate their own secretion. The evidence also suggests that at least part of the antigonadal effects of LHRH agonists given chronically could be ascribed to an inhibitory role of these peptides on endogenous LHRH secretion (Fig. 2).

D. Environmental Factors Affecting Gonadal Function

A number of chemical and physiological agents can damage the gonad and thereby induce serious or irreversible damage to the seminiferous epithelium. Different physical agents known to affect reproductive performance are heat or ionizing radiation. Irreversible damage to the stem cell of the germinal epithelium can occur with high or prolonged doses. In many cases, the damage is irreversible, and no effective therapy has been found to prevent or reverse this type of damage. Genetic damage may also occur, and this needs to be evaluated when lesions of these type are seen (see Chapter 9).

Concerning the effects of chemical agents, there is a growing list of potentially hazardous substances. Some are substances contaminating the environment, while others are taken for therapeutic purposes, that is, cytotoxic or chemotherapeutic agents. In either case, the potential to induce mutations in germ cells is of concern, as is the irreversibility of the testicular damage.

The effects of environmental factors upon the hypothalamic–pituitary axis are much less clear. In experimental animals, several neurotoxins, including monosodium glutamate, can produce direct lesion in hypothalamic nuclei, such as the arcuate nuclei, involved in the regulation of gonadotropin secretion. Early exposure to estrogens during the period of sexual differentiation of the brain can also produce permanent damage to the hypothalamic–pituitary–gonadal axis. Estrogens or substances with estrogenic activity can be derived from different sources in the environment, including plants, pesticides, animal feeds, etc. However, our knowledge of the contribution of these environmental factors to reproductive disorders based on hypothalamic–pituitary imbalance is still sketchy, and much more research is needed in this area to clarify these important issues.

III. Pituitary Factors Regulating Testicular Function

A. Luteinizing Hormone and Follicle-Stimulating Hormone

Both gonadotropins are complex protein, made up of two subunits, α and β. The LH α subunit is very similar if not identical to the α subunit of FSH, thyroid-stimulating hormone (TSH), and human chorionic gonadotropin (hCG). The β subunit of each molecule, therefore, confers the specificity for the biological activity of each of these proteins. Both LH and FSH are glycoproteins, containing the hexosamines *N*-acetylgalactosamine and *N*-acetylglycosamine, the hexoses mannose and α-galactose, and the methyl pentose fucose. Their biological functions in the testis involve stimulation of Leydig cells by LH, with the resulting enhancement of androgen production, and trophic effects on Sertoli cells by FSH (Fig. 3), which helps maintain the integrity of the spermatogenic epithelium. These actions of the gonadotropins will be discussed in detail in Chapter 4.

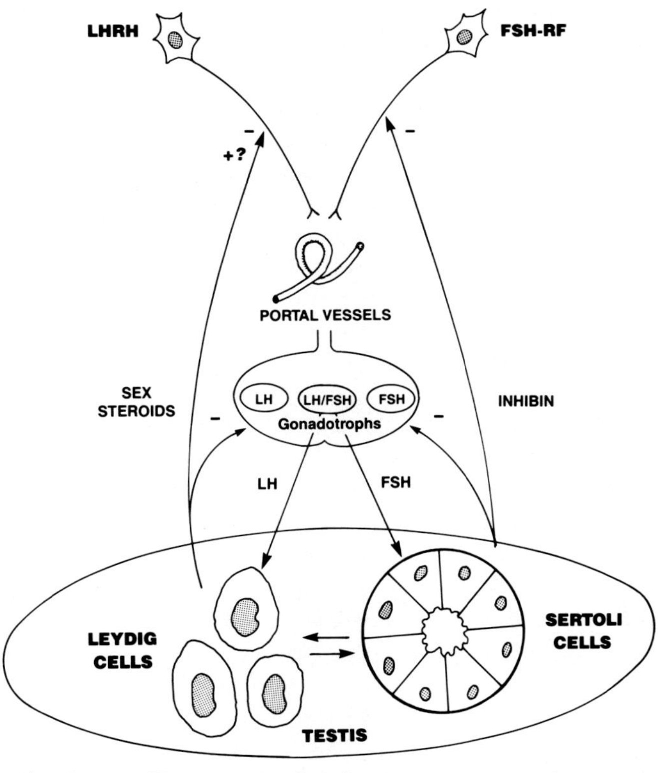

Fig. 3. This diagram illustrates the functional arrangement of the hypothalamic–pituitary–testicular axis. LHRH neurons release the decapeptide into the hypophyseal portal blood. At the pituitary level, LHRH activates specific receptors of similar or, possibly, different gonadotroph populations (e.g., LH-, FSH-, or LH/FSH-releasing cells), and as a result, the gonadotropins (LH and FSH) are secreted. In turn, LH binds to specific membrane receptors in the Leydig cells to promote the production of sex steroids, which, then, influence the secretion of LHRH and LH. On the other hand, putative FSH–RF neurons secrete such factor(s) that, in turn, would control the release of FSH from the gonadotrophs. By activating the receptors on Sertoli cells, FSH regulates the production and delivery of inhibin, which is a gonadal hormone that may inhibit FSH secretion directly at the level of the adenohypophysis and, in addition, through a central mechanism of action. Note that Sertoli cells synthesize and release factors that may modulate Leydig cell function and vice versa.

B. *Actions of Luteinizing Hormone–Releasing Hormone to Stimulate Luteinizing Hormone and Follicle-Stimulating Hormone Release*

As previously mentioned, a large body of evidence indicates that LHRH can stimulate both LH and FSH release by the pituitary, in animals and humans. *In vitro* experiments also show that LHRH stimulates not only release but also the synthesis of both gonadotropins. These observations resulted in the widely accepted view that LHRH is, in fact, a gonadotropin-releasing hormone (GnRH), the physiologic hypothalamic regulator of both LH and FSH release (Schally *et al.*, 1971). A number of situations exist, however, in which LH and FSH secretion are divergent. Examples include the late proestrus–early estrus surge of FSH alone, selective release or inhibition of FSH following stimulation or destruction of specific brain regions (most notably, the paraventricular nucleus/anterior hypothalamic area), differing rates of increase following gonadectomy, and differences in the pattern of pulsatile secretion (Negro-Vilar *et al.*, 1986). Despite these discrepancies, the concept of a single GnRH has survived and has been strengthened by studies that demonstrate that LHRH, administered in certain patterns and/or concentrations or during certain physiologic states, can selectively influence the secretion of either LH or FSH and alter the blood LH–FSH ratio. Elegant *in vivo* studies from Knobil's laboratory (Knobil, 1980) demonstrated that when rhesus monkeys with hypothalamic lesions placed to abolish endogenous LHRH secretion were given exogenous LHRH pulses, LH and FSH release could be concomitantly, antithetically, or selectively altered, depending on the frequency of the LHRH pulses. Recent studies analyzing pulsatile LH and FSH secretion have added new and significant information to the role of LHRH in the release of FSH (Culler and Negro-Vilar, 1986). Briefly, the data indicate that LHRH contributes significantly to the basal secretion of FSH. Removal of LHRH activity by passive immunoneutralization or specific receptor blockade with LHRH antagonists results in a clear decline in basal FSH levels. Pulsatile FSH secretion, however, remains unaltered. It is then necessary to postulate another level of regulation for pulsatile FSH secretion. This may involve a separate FSH-releasing hormone or, alternatively, some local factor in the pituitary that may regulate the pulsatile pattern of FSH release.

Numerous studies have been published describing the intracellular events that occur after LHRH binds to its specific receptor on the gonadotropin cell membrane. In a recent review, Conn (1986) describes in detail the molecular basis for LHRH action on the gonadotroph. Calcium has clearly been established as a second messenger for the action of LHRH. In the absence of Ca^{2+} or in the presence of ethylene glycol bis-(β-aminoethyl ether) *N*, *N*′-tetraacetic acid (EGTA), LHRH-induced LH release is completely suppressed even though binding of LHRH to its receptors still occurs. Calmodulin has been shown to be present in membrane patches containing the LHRH receptor, and it appears to be an intracellular receptor for Ca^{2+}, mobilized by LHRH stimulation. Inositol phospholipid breakdown has also been shown to be part of the early action of LHRH. Within seconds after LHRH stimulation, there is an immediate incorporation of [^{32}P] phosphate into phosphatidyl inositol (PI) and phosphatidic acid (PA), with subsequent, rapid changes in the pools of labeled PI, PI-4-phosphate, and PI-4,5-bisphosphate. Additional observations suggest that the turnover of inositol phospholipids in response to LHRH stimulation may be uncoupled from the Ca^{2+}–calmodulin-mediated pathway, stimulated by LHRH, to enhance LH release.

The activation of protein kinase C may also be involved in the LH secretory process. Both diacylglycerol and phorbol ester (Negro-Vilar and Lapetina, 1985) have been shown to stimulate LH secretion. The stimulatory activity of diacylglycerol and phorbol ester is time and concentration dependent and has been shown to include all pituitary hormones (Negro-Vilar and Lapetina, 1985), suggesting that this is a general mechanism for secretagogue-activated pituitary hormone secretion.

As is the case in many other hormonally controlled tissues, arachidonic acid and its metabolites play a role in the regulation of anterior pituitary hormone secretion. Early studies indicate that LHRH releases arachidonic acid from gonadotroph phospholipids. Moreover, inhibition of arachidonic acid release, by blocking phospholipase A2 activity, prevents LHRH-stimulated LH secretion (Naor and Catt, 1981). Arachidonic acid promotes LH release when added to pituitary cells *in vitro* (Naor and Catt, 1981; Snyder *et al.*, 1983). Early reports clearly established that prostaglandins are not involved in stimulating the release of LH. On the other hand, the lipoxygenase pathway has been reported to modulate, at least in part, the action of LHRH on pituitary gonadotrophs, and 5-HETE has been shown to be particularly active in promoting LH secretion. Blockade of lipoxygenase activity with NDGA or other blockers greatly reduces LHRH-stimulated LH secretion. Snyder *et al.*, (1983) also reported that 5,6-epoxyicosatrienoic acid, an arachidonate product derived from the NADPH-supported cytochrome P-450–dependent epoxygenase pathways, stimulated LH secretion with a potency at least equal to that of LHRH. These series of studies suggest that arachidonic acid and its lipoxygenase- and epoxygenase-derived metabolites might be a component of the cascade of events initiated by LHRH to promote LH secretion.

Several studies have eliminated a role for the major cyclic nucleotides (cAMP, cGMP) in the action of LHRH on gonadotropin release (Naor *et al.*,

1978; Conn *et al.*, 1979). This is consistent with the above observations, suggesting that the major, if not all, actions of LHRH are involved with Ca^{2+} mobilization and PI turnover and are independent of cAMP generation.

C. Prolactin Secretion and Role in Regulation of Gonadal Function

1. Effect of Steroids on Prolactin Secretion

Increasing evidence has accumulated during the past several years indicating that testosterone can stimulate prolactin secretion in rats, monkeys, and humans. Similarly, androstenedione has also been found to increase prolactin secretion. Since dihydrotestosterone, androstanediol, and androsterone have been found to have no stimulatory effect upon prolactin secretion, it has been suggested that only androgens that can be aromatized to estrogens (such as testosterone and androstenedione) are capable of increasing prolactin release (Negro-Vilar, 1980). Indeed, studies in juvenile rhesus monkeys show that testosterone administration can lead to increments in both serum estradiol and serum dihydrotestosterone. This indicates that aromatization of testosterone can produce increments in estradiol capable of stimulating prolactin. It remains to be established whether testosterone can stimulate prolactin secretion *per se* in the absence of aromatization to estradiol.

Estradiol and other natural and synthetic estrogens are the most potent hormonal stimulators of prolactin secretion in mammals of both sexes. In males, increments in prolactin secretion have been found after administration of estradiol to rats and of ethinyl estradiol to men. This stimulatory action of estradiol takes place at both pituitary and hypothalamic sites of actions, involving interactions with the dopaminergic and probably with other neurotransmitter systems. This potent stimulatory action of estradiol in prolactin secretion is sometimes overlooked in animal studies on the effect of estrogens on the prostate and seminal vesicles. Besides a direct effect of the steroid on the prostate, estrogens may also affect the accessory glands through changes in prolactin levels. These types of studies, therefore, should include hypophysectomized controls to evaluate the direct action of estradiol in the absence of other pituitary hormones (Negro-Vilar, 1980).

Several reports have been published suggesting a functional relationship between prolactin and the glucocorticoids. Adrenalectomy results in decreased serum levels of prolactin. Glucocorticoid injections in humans have been found to decrease serum prolactin levels. Conversely, prolactin seems to have a stimulatory role in the adrenal gland. High levels of serum prolactin resulted in increased adrenal weights in intact, castrated, or in hypophysectomized rats (Negro-Vilar *et al.*, 1977). It has also been found that prolactin increases corticosterone release from male adrenal glands, probably through an inhibition of adrenal 5α-reductase activity.

2. Effects of Prolactin on Male Reproductive Tract

The discovery of prolactin receptors in the testis and epididymis fostered the belief that prolactin is directly involved in the regulation of testicular function under normal conditions. Prolactin has been shown to stimulate testicular growth in hypophysectomized animals (Negro-Vilar *et al.*, 1977) and has also been reported to stimulate spermatogenesis under different experimental situations (Bartke *et al.*, 1978). Growth of the epididymis has been obtained by synergistic action of prolactin and dihydrotestosterone. At least some of the effects of prolactin on the testis seem to be exerted at the level of the Leydig cell (see Chapter 4), affecting androgen production and secretion. Prolactin was reported to act synergistically with LH to increase plasma testosterone, incorporation of labeled precursors to testosterone, and testosterone output *in vitro*. Labeled prolactin has been localized by autoradiography in mouse Leydig cells. Specific binding of prolactin to isolated Leydig cells has also been shown to occur. A physiological involvement of prolactin in Leydig cell function in men is also supported by observations correlating peripheral prolactin and testosterone levels. Additional experimental evidence indicates that prolactin may physiologically increase testicular sensitivity to LH during the course of sexual maturation. These observations indicate that normal circulating levels of prolactin would stimulate steroidogenesis through an effect in the Leydig cell acting alone and/or synergistically with LH. On the other hand, high serum prolactin levels, as seen in hyperprolactinemic patients, may in effect be inhibitory to testicular function. Hyperprolactinemic males have been reported to have smaller testes and prostates and decreased volume of the seminal ejaculate and are in some cases impotent (Thorner *et al.*, 1977).

D. Pulsatile Pituitary Hormone Secretion

1. The Significance of Pulsatile Hormone Secretion to Optimal Reproductive Capacity

Much evidence has accumulated during the last decade to indicate that most hormones are secreted from their respective cells, into the circulation, in a pulsatile fashion (Negro-Vilar *et al.*, 1986). This pattern of secretion is normally controlled by the central nervous system, although some evidence suggests that the isolated organs can also release hormones in a pulsatile manner. The significance of these observations has been clarified by studies that indicate that when secretagogues are infused continuously or in various pulsatile patterns *in vivo,* the response of the target organ can vary dramatically, ranging from normal or elevated responses to decreased or suppressed secretion.

Perhaps the best example of the impact that the pattern of stimulation can have on the function of a system has been provided by the work from Knobil's laboratory examining the role of exogenously administered LHRH on pituitary gonadotropin secretion patterns in rhesus monkeys, whose endogenous input signal (LHRH) had been suppressed by mediobasal hypothalamic lesion. In such animals, LH and FSH secretion fall to very low or undetectable levels, but can be reestablished by the pulsatile administration of LHRH. Conversely, if the pulsatile administration of LHRH is replaced by a continuous i.v. infusion, an inhibition of gonadotropin secretion is observed that persists throughout the infusion period. When the pulsatile mode of LHRH administration is reestablished, secretion of both gonadotropins increases dramatically. In addition, it has been demonstrated that not only is the type of stimulation pattern (pulsatile versus constant) important in determining the gonadotropin release response but also the characteristics of the pulsatile pattern. For example, changing the frequency of the LHRH pulses can subsequently alter the release of LH and FSH. Decreasing the frequency of the LHRH pulses from one per hour to one pulse every 3 h dramatically reduces LH secretion while producing a modest increase in FSH release, resulting in an increased plasma FSH – LH ratio. Restoring the frequency to one pulse per hour returns both hormones to normal plasma levels and ratio. These and other data suggest that the absolute amount of peptide infused contributes little to the response obtained. The most relevant component of the input signal is the pattern of administration of the hormone. These observations are of fundamental importance for fertility and infertility therapies and diagnosis. Knowledge of the appropriate pattern of administration of LHRH has allowed the establishment of therapeutic pulsatile LHRH treatment in males with hypothalamic hypogonadism. Analysis of potential disturbances in the normal pattern of pulsatile hormone secretion can lead to a better understanding of the various disorders now classified as hypothalamic – pituitary deficiencies.

2. Hypothalamic Factors Regulating Pulsatile Gonadotropin Secretion

Early studies indicate that LH is secreted in a pulsatile or episodic fashion in many laboratory animals and in humans (Knobil, 1980). Additional studies clearly show that both gonadotropins and LHRH are secreted in an episodic fashion (Negro-Vilar *et al.*, 1986). Further evidence presented by Knobil (1980), as discussed above, indicates that pulsatile LHRH release is essential for sustained stimulation of LH release. Those studies provided the experimental and rational basis for the therapeutic application of pulsatile LHRH substitution in the treatment of hypothalamic hypogonadism (Crowley and McArthur, 1980).

Evidence from our laboratory indicates that each episode of LH release is preceded by the release of LHRH from the ME. Simultaneous analysis of dopamine and LHRH levels in three different preoptico–hypothalamic regions was performed and correlated with the LH secretory episode. For this purpose frequent blood samples were obtained from chronically cannulated, freely moving, ovariectomized rats during a 2-h period. In this way it was possible to reconstruct in every case the relative position in the LH secretory episode at the time that all amine and peptide determinations were performed (Negro-Vilar *et al.*, 1982). The results show that prior to the rising phase in each LH secretory episode there is a concomitant drop in dopamine and LHRH levels in the ME, suggesting that both dopamine and LHRH are released prior to, or in conjunction with, the initiation of the LH surge. Concomitantly with these changes, a parallel rise in NE levels in the suprachiasmatic–medial preoptic region occurred, which we interpreted as suggestive of increased noradrenergic activity in the area at the time of the LH surge. The following comprehensive hypothesis can be formulated to account for the sequence of events leading to pulsatile LH release: noradrenergic input originating in the midbrain and projecting to the suprachiasmatic–medial preoptic region activates LHRH perikarya located in that region, resulting in the transport of peptide to more caudal and basal hypothalamic areas. This provides a continuous replenishment of LHRH stores in the ME to replace the peptide that is periodically released from that region. Further noradrenergic input to the basal hypothalamic areas may also play a role in LHRH release. In addition, dopamine neurons within the tuberoinfundibular system projecting to the ME stimulate LHRH release. The oscillator that controls the episodic release of LHRH is entrained to appropriate environmental and internal cues that modify the frequency and characteristics of the pulses. It is obvious that such a delicate control system can be affected by a number of exogenous and endogenous factors, and the evidence accumulating so far suggests that indeed different neurotoxins, steroids, and other chemical agents can derange the activity of the hypothalamic pulse generator and affect thereby the gonadotropin control of testicular function.

The FSH, like LH, is secreted in a pulsatile manner in both male and female gonadectomized rats (Negro-Vilar *et al.*, 1986). Pulse frequency of the two gonadotropins is similar in rats of both sexes: approximately one LH pulse every 30 min and one FSH pulse every 40 min. Pulsatile LH and pulsatile FSH secretion are believed to be regulated by episodic release of

LHRH from the hypothalamus. However, since pulse frequency is believed to be an index of hypothalamic input and pulse amplitude an index of pituitary response, it can be argued that divergent hypothalamic releasing mechanisms for LH and FSH must exist on the basis of the observed difference in pulse frequency. Further differences in pulsatile LH and FSH secretion have been observed with regard to the response to feedback effects of ovarian factors. Significant reductions in both FSH pulse frequency and amplitude were seen in ovariectomized rats injected with steroid-stripped porcine follicular fluid, a source of inhibin, while pulsatile LH secretion was unaffected. Conversely, injection of estradiol benzoate significantly lowered LH pulse frequency and the mean plasma levels of both LH and FSH, without altering FSH pulse frequency or amplitude. Interestingly, when the porcine follicular fluid preparation and estradiol benzoate treatments are given in combination, all parameters of both LH and FSH secretion are significantly reduced. While it can be argued that the divergent LH and FSH responses are due to selective alteration in pituitary sensitivity and/or a change in the concentration of LHRH released from the hypothalamus that favors one or the other gonadotropin, the selective changes in pulse frequencies also argue in favor of divergent hypothalamic releasing mechanisms. This possibility is further supported by the knowledge that both estradiol and porcine follicular fluid (inhibin) have hypothalamic sites of action, as well as direct effects at the level of the pituitary (Negro-Vilar and Lumpkin, 1983).

Recent studies in orchidectomized rats, designed to obliterate endogenous LHRH secretion, have revealed striking differences in the relative dependence of LH and FSH secretion on LHRH. Endogenous LHRH was immunoneutralized by intravenous injection of a potent LHRH antiserum. Frequent blood sampling at different intervals after this treatment indicates that an abrupt interruption in the LH pulsatile pattern occurs immediately after administration of the antiserum. Mean plasma LH levels also drop to levels comparable to those seen in intact animals. Pulsatile FSH secretion, on the other hand, was unaffected by the treatment. Baseline FSH levels, however, declined by 24 and 48 h by about 50%. Replacement therapy, using an LHRH agonist not recognized by the antiserum, restored basal LH and FSH levels to pretreatment values, while not affecting pulsatile FSH release. These observations clearly indicate that (a) LH release is entirely dependent on sustained LHRH output, both in its tonic and pulsatile pattern; (b) basal FSH release is not as tightly controlled by LHRH as is that of LH, since changes in basal levels after LHRH blockade occur more sluggishly and replacement therapy also requires several hours to restore basal levels to pretreatment values; and (c) pulsatile FSH release is independent of LHRH and may be under the control of separate hypothalamic and/or pituitary factors.

IV. Reproductive Dysfunction and Guidelines for Endocrine Exploration of the Infertile Male

The following concepts are designed to serve as a guideline for the toxicologist who is faced (a) with the evaluation of a drug or compound that causes fertility problems or (b) with the design of a series of tests to determine potential adverse reproductive effects of a given substance. The approach recommended is to take a series of steps to determine the nature, site, mechanism, and possible causes of the problem. These recommendations are not designed to meet the requirements of regulatory agencies but rather to provide the toxicologist with some simple guidelines to obtain some essential information about the potential impact of a given drug or chemical on the reproductive performance of animals or humans, and in particular to explore the endocrine hypothalamic–pituitary–gonadal axis.

A. Target Sites for Toxicity

The preceding sections have provided the reader with an idea of the many intricacies and complexities of the mechanisms that regulate the activity of the hypothalamic–pituitary–testicular axis. It follows, then, that the sites and mechanisms by which different chemicals or drugs can interfere with reproductive functions are manyfold. Indeed, a complete analysis and exploration of the actions of a given substance may pose complex problems and require a long and complicated series of studies, perhaps developing into a lengthy, extensive, and costly process. This may fall into the realm of toxicological research, and, if so, the pathway to be taken and the extent of the exploration need only be limited by the interest of the researcher. On the other hand, practical considerations may indicate a series of shorter, more diagnostic tests, designed primarily to ascertain whether a compound has toxic effects on the reproductive sphere, what kind of effects, and what site(s) is the primary target.

A simple division of potential target sites for toxicants can be based on both anatomical as well as functional aspects. An initial classification will divide the potential origin of the problem among (a) hypothalamic or central disorders, (b) pituitary disorders, and (c) testicular and/or accessory gland defects.

Once the problem is circumscribed to one of these three main sites, further subdivision is possible. For instance, a testicular site may compromise the interstitial (Leydig cell), vascular, support (Sertoli cell), or germinal compartment, or a combination of two or more of the above. Appropriate design of different tests may provide clear data to arrive at a reasonably clear diagnosis of the site and nature of the problem.

B. Exploratory Approach

The first question to be asked is: does the compound in question have a known effect on reproductive performance? If the compound has never been tested, then a series of short-term tests should be performed to ascertain the potential toxic hazard on the reproductive area, as indicated below.

1. Nature of the Problem

Once the pharmacokinetic and pharmacodynamic properties of the substance to be tested are determined, short-term (3–6 months) toxicity studies should be conducted, with doses of the substance that are derived from the pharmacokinetic profile of the drug. In the exploration of the male, the usual pathology performed at the end of standard toxicological tests is, in our opinion, insufficient to provide adequate information about the potential hazard of a chemical. Many functional defects, in particular at the CNS level, do not present a clear anatomical substrate. Therefore, standard pathology evaluation is not going to provide the required information. Perhaps the most practical approach to detect problems of reproductive performance is to conduct a two- to three-generation study in mice or rats (see Chapter 6). This type of test provides very useful information not only about reproductive performance of the male but also about other possible mutagenic, developmental, and behavioral defects.

Once a decline in fertility is detected, the next question should be directed to determine what component of the reproductive cycle has been compromised. Is there a behavioral (copulatory) deficit, or an endocrine disorder, or a failure of the germinal epithelium, etc.?

2. Site of the Problem

An analysis of the data obtained from the initial test may already shed some light on the level of the toxic lesion. However, it is important to clearly ascertain the site(s) of the lesion, because this may indicate the potential cause of the problem and will also suggest the type of tests that are necessary as a follow-up to establish the correct diagnosis. These tests are necessary because, for example, a testicular lesion need not be interpreted as resulting from a direct effect of the compound on the testis. A number of substances with neurotoxic potential may compromise the hypothalamic–pituitary axis with a resulting loss of gonadotropic stimulation of the testis. This will lead to a decline in both the endocrine and the spermatogenic function of the testis, and in a chronic situation will result in clear damage to the gonad, which will be detected in the anatomo–pathological examination. Therefore, the diagnostic test that may help determine the causes of the problem will also help resolve the site(s) and nature of the problem.

3. Possible Causes of the Problem

As a complement to the pathology performed at the end of the test, a full endocrine and biochemical evaluation of blood and other fluid samples should be performed *during* as well as at the termination of the test treatment period. In many cases, blood chemistry and hormonal measurements are performed at the time of sacrifice. This is usually not informative enough. Taking the example of circulating pituitary and gonadal hormones, we know very well now, as discussed earlier, that all hormones are secreted in a pulsatile pattern, with relatively wide fluctuation in plasma values between nadir (trough) and peak of each pulse. These pulsatile variations are superimposed on circadian, as well as seasonal, rhythms, which also contribute to a wide fluctuation in plasma hormone levels. Therefore a single point determination of LH, FSH, or testosterone levels, for example, may yield very little, if any, useful information. To say that a hormone value is "within normal limits" is of no significance if the range of variation in the hormone is too wide. As we indicated above, pulse frequency is perhaps the most important factor in the hormonal secretory process. Unless serial samples are obtained, a marked change in pulse frequency, or even the absence of pulses, may remain undetected if the mean value of the hormone is "within normal limits."

Practical or cost considerations may impede the performance of elaborate tests of hormone secretory patterns as described above. In lieu of these tests, alternative experiments could be performed in which the secretory capacity of a given organ (hypothalamus, pituitary, or testis) or cell type (gonadotroph, Leydig cell, etc.) is tested with a specific secretagogue. This requires that doses of the secretagogue employed are within the range of the physiological response and, if possible, performed at multiple doses within the dose–response curve. Use of very large doses of hCG, for example, to stimulate testosterone secretion may obscure a deficit in the response to doses of the hormone within the physiological range.

4. Basic Test for Evaluation of Reproductive Toxicity: *In Vitro* and *In Vivo* Approaches

The following recommendations should be considered as a basic guideline of simple tests designed to ascertain the potential site(s) of action of an agent and the degree of compromise of the functional response integrated at a given level.

To test for hypothalamic dysfunction a combination of *in vivo* and *in vitro* approaches could be used. A useful test of hypothalamic responsiveness is to test the response to orchidectomy at different intervals. Within 10–12 h after removal of the testes, the male rat responds with a significant increase in LH and FSH secretion. This rise in the gonadotropins continues for several weeks. In cases of severe hypothalamic dysfunction, a sluggish or delayed rise in the gonadotropin will be observed. If the hypothalamic lesion is very marked, no increase in gonadotropins may be observed. A lack of change in the rise of the gonadotropins, on the other hand, does not completely rule out a hypothalamic dysfunction, perhaps involving more subtle changes in the regulation of pituitary hormone secretion. In this case, an analysis of pulsatile hormone secretion, looking at pulse frequency, amplitude, etc., may provide some positive results.

If the substance tested has intrinsic steroidal activity, the rise in gonadotropins may be blocked not because of hypothalamic lesion but because of the negative feedback effect of the drug at both hypothalamic and pituitary sites.

Additional tests of hypothalamic function may include studies on LHRH release *in vitro* from dissected hypothalamic fragments, evaluation of LHRH levels and distribution in brain by immunocytochemistry and radioimmunoassay, etc. Recently, some excitatory amino acids such as aspartate and other analogs have been used in rats and rhesus monkeys after i.v. administration to test the capacity of the hypothalamus to release LHRH. The latter is evaluated indirectly by the increase in LH in peripheral plasma.

Finally, hypothalamic and extrahypothalamic lesions may affect fertility by modifying sex behavior. Standard tests of copulatory behavior can be performed to measure these parameters.

Pituitary function can more easily be explored *in vivo* by testing for possible changes in sensitivity to LHRH. It is important to distinguish between sensitivity and responsiveness. The former requires to find a dose of the secretagogue that is not effective and several doses over the range of the physiological response. Responsiveness can be evaluated by using one or two doses known to be within the physiological response. In either case, an appropriate time course to follow up the LH and FSH response should be included. *In vitro* studies employing anterior pituitary fragments or cultured, dispersed pituitary cells can be used to evaluate whether the drug being tested can directly modify pituitary hormone secretion and/or interfere with the ability of known secretagogues (like LHRH) to stimulate hormone release.

Testicular endocrine function can easily be tested *in vivo* by measuring resting testosterone levels as well as the response to the administration of different doses of LH or hCG. Again, time courses and doses should be optimized to ensure that a defect, if present, can be detected. *In vitro* incubations of decapsulated, whole testis or of dispersed, purified Leydig cells can also be used to evaluate direct and/or interfering effects of the substance on testosterone production.

Isolation of these three levels—hypothalamic, pituitary, and testicular—

should provide useful, direct information about the site(s) of action of an agent interfering with the endocrine regulation of testicular gametogenesis and reproductive performance. Damage to the germinal cell line or other support cell systems (i.e., Sertoli cell) in the testis requires a different approach, as discussed in other chapters (2 and 10) of this volume. More sophisticated experimental approaches may then be used to determine the intimate cellular and subcellular mechanisms mediating the toxic effects of chemicals on the reproductive system.

References

Adelman, J. P., Mason, A. J., Hayflick, J. S., and Seeburg, P. H. (1986). Isolation of the gene and hypothalamic cDNA for common precursor of gonadotropin-releasing hormone and prolactin release-inhibiting factor in human and rat. *Proc. Natl. Acad. Sci. USA* **83,** 179–183.

Advis, J. P., McCann, S. M., and Negro-Vilar, A. (1980). Evidence that catecholaminergic and peptidergic (LHRH) neurons insuprechiasmatic-medial preoptic medial basal hypothalamus and median eminence are involved in estrogen negative feedback. *Endocrinology* **107,** 892–901.

Anthony, E. L. P., King J. C., and Stopa, E. G. (1984). Immunocytochemical localization of LHRH in the median eminence, infundibular stalk, and neurohypophysis. *Cell Tissue Res.* **236,** 5–14.

Arimura, A. (1976). Recent developments in the study of hypothalamic hormones with special references to LHRH and somatostatin. *Nippon Naibumpi Gakkai Zasshi* **52,** 1159–1183.

Baja, J. S., and Madan, R. (1983). New approaches to male fertility regulation: LHRH analogs, steroidal contraception, and inhibin. *In* "Endocrine Mechanisms in Fertility Regulation" (G. Benagiano and E. Diczfalusy, eds.), pp. 203–231. Raven, New York.

Bartke, A. (1986). Hyperprolactinemia and male reproduction. *In* "Andrology: Male Fertility and Sterility" (J. Paulson, A. Negro-Vilar, E. Lucena, and L. Martine, eds.), pp. 101–123. Academic Press, Orlando, Florida.

Bartke, A., Hafiez, A. A., Bex, F. J., and Dalterio, S. (1978). Hormonal interactions in regulation of androgen secretion. *Biol. Reprod.* **18,** 44–54.

Bennett, G. H., Edwardson, J. A., Holland, D., Jeffcoate, S. L., and White, N. (1975). Release of immunoreactive luteinizing hormone-releasing hormone and thyrotropin-releasing hormone from hypothalamic synaptosomes. *Nature (London)* **257,** 323–325.

Bhasin, R., Heber, D., Peterson, M., and Swerdloff, R. (1983). Partial isolation and characterization of testicular GnRH-like factors. *Endocrinology* **112,** 1144.

Bjorklund, A., Lindvall, O., and Nobin A. (1975). Evidence of an incerto-hypothalamic dopamine neuron system in the rat. *Brain Res.* **89,** 29.

Brown-Grant, K. (1976). Control of gonadotropin secretion. *In* "Subcellular Mechanism in Reproductive Neuroendocrinology" (F. Naftolin, K. J. Ryan, and I. J. Davies, eds.), pp. 485–502. Elsevier, Amsterdam.

Brownstein, M. J., Saavedra, J. M., Palkovits, M., and Axelrod, J. (1974). Histamine content of hypothalamic nuclei of the rat. *Brain Res.* **77,** 151.

Chiocchio, S. R., Negro-Vilar, A., and Tramezzani, J. H. (1976). Acute changes in norepinephrine content in the median eminence induced by orchidectomy or testosterone replacement. *Endocrinology* **99,** 629.

Clayton, R. N., and Catt, K. J. (1980). Receptor-binding affinity of gonadotropin-releasing hormone analogs: Analysis by radioligand receptor assay. *Endocrinology* **106,** 1154.

Clemens, J. A., Tinsley, F. C., and Fuller, R. W. (1977). Evidence for a dopaminergic component in the series of neural events that lead to the proestrous surge of LH. *Acta Endocrinol.* **85,** 18–24.

Collu, R., Fraschini, F., and Martini, L. (1973). Role of indoleamines and catecholamines in the control of gonadotropin and growth hormone secretion. *Prog. Brain Res.* **39,** 289–299.

Conn, P. M. (1986). The molecular basis of gonadotropin releasing hormone action. *Endocr. Rev.* **7,** 3–10.

Conn, P. M., Dufau, M. L., and Catt, K. J. (1979). GnRH-stimulated release of LH from rat pituicytes does not require production of cyclic AMP. *Endocrinology* **104,** 448–453.

Crowley, W. F., and McArthur, J. W. (1980). Stimulation of the normal menstrual cycle in Kallman's syndrome by pulsatile administration of luteinizing hormone hormone–releasing hormone (LHRH). *J. Clin. Endocrinol. Metab.* **51,** 173.

Culler, M. D., and Negro-Vilar, A. (1986a). Development of specific antisera and a radioimmunossay procedure for the gonadotropin-releasing hormone associated peptide (GAP) of the LHRH-prohormone. *Brain Res. Bull.* **17,** 219–223.

Culler, M. D., and Negro-Vilar, A. (1986b). Evidence that pulsatile follicle-stimulating hormone secretion is independent of endogenous luteinizing hormone–releasing hormone. *Endocrinology* **118,** 609–612.

Davidson, J. M. (1969). Feedback control of gonadotropin secretion. *In* "Frontiers in Neuroendocrinology" (W. F. Ganong and L. Martini, eds.), pp. 343–388. Academic Press, New York.

Davidson, J. M., Damassa, D. A., Smith, E. R., and Cheung, C. (1976). Feedback control of gonadotropin secretion in the rat. *In* "Regulatory Mechanisms of Male Reproductive Physiology" (C. H. Spilman, T. J. Lobol, and K. T. Kirton, eds.), pp. 151–168. Excerpta Medica, Amsterdam.

Drouva, S. V., and Gallo, R. V. (1976). Catecholamine involvement in episodicluteinizing hormone release in adult ovariectomized rats. *Endocrinology* **99,** 651–658.

Fink, G. (1979). Feedback actions of target hormones on hypothalamus and pituitary, with special reference to gonadal steroids. *Annu. Rev. Physiol.* **41,** 571.

Fiorindo, R., Justo, G., Motta, M., Simonovic, I., and Martini L. (1974). *In* "Hypothalamic Hormones" (M. Motta, P. G. Crosignani, and L. Martini, eds.), pp. 195–204. Academic Press, New York.

Franchimont, P., Verstraelen-Proyard, J., Hazee-Hagelstein, M. T., Renard, Ch., Demoulin, A., Bourguignon, J. P., and Hustin, H. (1979). Inhibin: From concept to reality. *Vitam. Horm.* **37,** 243.

Freeman, M. E., Dupke, K. C., and Croteau, C. M. (1976). Extinction of the estrogen-induced daily signal for LH release in the rat: A role for the proestrous surge of progesterone. *Endocrinology* **99,** 223.

Gallardo, E. A., Voloschin, L. M., and Negro-Vilar, A. (1978). Effect of deafferentation of the medial basal hypothalamus on catecholamine content of the median eminence and on serum LH and prolactin levels in ovariectomized rats. *Brain Res.* **148,** 121–127.

Gallo, R. V., and Moberg, G. P. (1977). Serotonin mediated inhibition of luteinizing hormone release during electrical stimulation of the arcuate nucleus in ovariectomized rat. *Endocrinology* **100,** 945–954.

Gnodde, H. P., and Schuiling, G. A. (1976). Involvement of catecholaminergic and cholinergic mechanisms in the pulsatile release of LH in the long-term ovariectomized rat. *Neuroendocrinology* **20,** 212–223.

Héry, M., Laplante, E., and Kordon, C. (1976). Participation of serotonin in the phasic release of LH: I. Evidence from pharmacological experiments. *Endocrinology* **99,** 496–503.

Hokfelt, T., Fuxe, K., Goldstein, M., and Johansson, O. (1974). Immunohistochemical evidence for the existence of adrenaline neurons in the rat brain. *Brain Res.* **66,** 235.

Kalra, S. P., and Kalra, P. S. (1984). Opioid–adrenergic–steroid connection in regulation of luteinizing hormone secretion in the rat. *Neuroendocrinology* **38,** 418–426.

Kalra, S. P., and McCann, S. M. (1974). Effect of drugs modifying catecholamine synthesis on plasma LH and ovulation in the rat. *Neuroendocrinology* **15,** 79.

Karsch, F. J., Dierschke, D. J., Weick, R. F., Yamaji, T., Hotchkiss, J., and Knobil, E. (1973). Positive and negative feedback control by estrogen of luteinizing hormone secretion in the rhesus monkey. *Endocrinology* **92,** 799.

King, J. C., and Anthony, E. L. P. (1983). Biosynthesis of LHRH: inferences from immunocytochemical studies. *Peptides* **4,** 963–970.

King, J. C., Tobet, S. A., Snavaly, F. L., and Arimura, A. A. (1982). LHRH immunopositive cells and their projections to the median eminence and organum vasculosum of the lamina terminalis. *J. Comp. Neurol.* **290,** 287–300.

Knobil, E. (1980). The neuroendocrine control of the menstrual cycle. In "Recent Progress in Hormone Research" (R. O. Greep, ed.), pp. 53–88. Academic Press, New York.

Krieg, R. J., and Sawyer, C. H. (1976). Effects of intraventricular catecholamines on luteinizing hormone release in ovariectomized-steroid primed rats. *Endocrinology* **99,** 411.

Leranth, C., MacLusky, N. J., Sakamoto, H., Shanabrough, M., and Naftolin, F. (1985). Glutamic acid decarboxylase-containing axons synapse on LHRH neurons in the rat medial preoptic area. *Neuroendocrinology* **40,** 536–539.

Ling, N. Ying, S. Y., Ueno, N., Shimasaki, S., Esch, F., Hotta, M., and Guillemin, R. (1986). Pituitary FSH is released by a heterodymer of the β subunits from the two forms of inhibin. *Nature (London)* **321,** 779–782.

Liposits, Z., Setalo, G., and Flerko, B. (1984). Application of the silver–gold intensifier 3,3′—diaminobenzidine chromogen to the light and electron microscopic detection of the luteinizing hormone–releasing hormone system of the rat brain. *Neuroscience* **13,** 513–525.

Lofstrom, A. (1977). Catecholamine turnover alterations in discrete areas of the median eminence of the 4- and 5-days cyclic rats. *Brain Res.* **120,** 113.

Lumpkin, M., Negro-Vilar, A., Franchimont, P., and McCann, S. (1981). Evidence for a hypothalamic site of action of inhibin to suppress FSH release. *Endocrinology* **108,** 1101.

McCullagh, D. R. (1932). Dual endocrine activity of testis. *Science* **76,** 19–20.

Marshall, J. C., Bourne, G. A., Frager, M. S., and Pieper, D. R. (1981). Pituitary gonadotropin–releasing hormone receptors-physiological changes and control of receptor number. *In* "Functional Correlates of Hormone Receptors in Reproduction" (V. Mahesh, T. Muldoon, B. L. Saxena, and W. Sadler, eds.), pp. 93–115. Elsevier/North-Holland, New York.

Marshall, P. E., and Goldsmith, P. C. (1980). Neuroregulatory and neuroendocrine GnRH pathways in the hypothalamus and forebrain of the baboon. *Brain Res.* **193,** 353.

Martinovic, J. V., and McCann, S. M. (1977). Effect of lesions in the ventral noradrenergic tract produced by microinjections of 6-hydroxydopamine on gonadotropin release in the rat. *Endocrinology* **100,** 1206–1213.

Mason, A. J., Hayflick, J. S., Ling, N., Esch, F., Ueno, N., Ying, S. Y., Guillemin, R., Niall, H., and Seeburg, P. H. (1985). Complementary DNA sequences of ovarian follicular fluid inhibin show precursor structure and homology with transforming growth factor β. *Nature (London)* **318,** 659–663.

Masotto, C., and Negro-Vilar, A. (1986). GABA and gonatropin secretion: Evidence from *in vitro* studies on regulation of LHRH secretion. In "GABA and Endocrine Function" (G. Racagni and A. O. Donoso, eds.), pp. 243–250. Raven, New York.

Meites, J., Van Vugt, D. A., Forman, L. J., Syvester, P. W., Ieire, T., and Sonntag, W. (1983). Evidence that endogenous opiates are involved in control of gonadotropin secretion. *In* "The Anterior Pituitary Gland" (A. S. Bhatnager, ed.), pp. 327–340. Raven, New York.

Millar, R. P., Wormald, P. J., and Milton, R. C. de L. (1986). Stimulation of gonadotropin releasing by a non-GnRH peptide sequence of the GnRH precursor. *Science* **232,** 68–70.

Muller, E. E., Cocchi, D., Locatelli, V., Apud, J. A., Tappaz, M. L., Masotto, C., Novelli, A., and Racagni, G. (1983). GABA receptor regulation of endocrine function. *In* "The GABA Receptor" (S. J. Enna, ed.), pp. 257–304. Humana, New York.

Naor, Z., and Catt, K. J. (1981). Mechanism of action of gonadotropin releasing hormone, involvement of phospholipid turnover in luteinizing hormone release. *J. Biol. Chem.* **256,** 2226–2229.

Naor, Z., Zor, V., Meidan, R., and Koch, Y. (1978). Sex differences in pituitary cyclic AMP response to gonadotropin releasing hormone. *Am. J. Physiol.* **235,** E37–E41.

Negro-Vilar, A. (1973). Interaction between gonadal steroids and LH-releasing hormone to control gonadotropin secretion at the pituitary level. *Acta Physiol. Lat. Am.* **23,** 494.

Negro-Vilar, A. (1980). Prolactin and the growth of prostate and seminal vesicles. *In* "Male Accessory Sex Glands" (E. Spring-Mills and E. S. E. Hafez, eds.), pp. 223–233. Elsevier Biomed., Amsterdam.

Negro-Vilar, A., and DePaolo, L. V. (1981). Brain–gonad interactions in the regulation of gonadotropin and prolactin release. *In* "Oligozoospermia: Recent Progress in Andrology" (G. Frajese, E. S. E. Hafez, C. Conti, and A. Fabbrini, eds.), pp. 323–336. Raven, New York.

Negro-Vilar, A. and Lapetina, E. (1985). 1,2-Didecanoylglycerol and phorbol 12,13-dibutyrate enhance anterior pituitary hormone secretion *in vitro. Endocrinology* **117,** 1559–1564.

Negro-Vilar, A., and Lumpkin, M. D. (1983). Inhibin: Central and peripheral effects to regulate follicle stimulating hormone secretion. *In* "Male Reproduction and Fertility" (A. Negro-Vilar, ed.), pp. 159–169. Raven, New York.

Negro-Vilar, A., and Valenca, M. M. (1985). Opiatergic regulation of LHRH release is mediated by catecholamines and prostaglandin E2. *Neural Endocr. Peptides Recept. Spring Symp.,* Washington, D.C., abstr. 36.

Negro-Vilar, A., Chiocchio, S. R., and Tramezzani, J. H. (1977a). Changes in catecholamine content of the median eminence precede the proestrous surges of luteinizing hormone and prolactin. *J. Endocrinol.* **75,** 339.

Negro-Vilar, A., Saad, W. A., and McCann, S. M. (1977b). Evidence for a role of prolactin secretion in prostate and seminal vesicle growth in immature male rats. *Endocrinology* **100,** 729–737.

Negro-Vilar, A., Ojeda, S. R., and McCann, S. M. (1979). Catecholaminergic modulation of LHRH release by median eminence terminals *in vitro. Endocrinology* **104,** 1749–1749.

Negro-Vilar, A., Advis, J. P., Ojeda, S. R., and McCann, S. M. (1982). Pulsatile LH patterns in ovariectomized (OVX) rat: Involvement of norepinephrine and dopamine in the release of LHRH and LH. *Endocrinology* **11,** 932–937.

Negro-Vilar, A., Valenca, M. M., and Masotto, C. (1985). Tonic inhibitory effect of endogenous opiates on gonadotropin secretion: Site and Mechanism of action. *J. Androl.* **6,** 82.

Negro-Vilar, A., Culler, M. D., and Masotto, C. (1986). Peptide–steroid interactions in brain regulation of pulsatile gonadotropin secretion. *J. Steroid Biochem.* **25,** 741–747.

Nikolics, K., Mason, A. J., Szonyi, E., Ramachandran, J., and Seeburg, P. H. (1985) A prolactin-inhibiting factor within the precursor for human gonadotropin-releasing hormone. *Nature (London)* **316,** 511–517.

Ojeda, S. R., and McCann, S. M. (1973). Evidence for participation of a catecholaminergic mechanism in the postcastration rise in plasma gonadotropins. *Neuroendocrinology* **12,** 295.

Ojeda, S. R., Negro-Vilar, A., and McCann, S. M. (1982). Evidence for involvement of α-adrenergic receptors in norepinephrine-induced prostaglandin E2 and luteinizing hormone–releasing hormone release from the median eminence. *Endocrinology* **110,** 409–412.

Palkovits, M. (1978). Topography of chemically identified neurons in the central nervous system:a review. *Acta Morphol. Acad. Sci. Hung.* **26,** 211–290.

Rotsztejn, W. H., Charli, J. L., Pattou, E., Epelbaum, J., and White, N. (1976). *In vitro* release of luteinizing hormone releasing hormone (LHRH) from rat mediobasal hypothalamus: Effects of potassium, calcium and dopamine. *Endocrinology* **99,** 1663–1666.

Saavedra, J. M., Palkovits, M., Brownstein, M. J., and Axelrod, J. (1974). Serotonin distribution in the nuclei of the rat hypothalamus and preoptic region. *Brain Res.* **77,** 157.

Sandow, J. (1986). Therapeutic use of LHRH analogs in reproductive disorders. *In* "Andrology: Male Fertility and Sterility" (J. Paulson, A. Negro-Vilar, E. Lucena, and L. Martini, eds.), pp. 15–37. Academic Press, Orlando, Florida.

Santen, R. J. (1977). Independent effects of testosterone and estradiol on the secretion of gonadotropin in man. *In* "The Testis in Normal and Infertile Men" (P. Troen, and H. R. Nankin, eds.), pp. 197–210. Raven, New York.

Sar, M., Culler M. D., McGimsey W. C., and Negro-Vilar, A. (1987). Immunocytochemical localization of the gonadotropin-releasing hormone-associated peptide of the LHRH precursor. *Neuroendocrinology* **45,** 172–175.

Schally, A. V., Kastin, A. J., and Arimura, A. (1971). Hypothalamic follicle-stimulating hormone (FSH) and luteinizing hormone (LH) regulating hormone: Structure, physiology and clinical studies. *Fertil. Steril.* **22,** 703–721.

Schally, A. V., Coy, D. H., and Meyers, C. A. (1978). Hypothalamic regulatory hormones. *Annu. Rev. Biochem.* **47,** 89.

Setalo, G., Vigh, S., Schally, A. V., Arimura, A., and Flerko, B. (1976). Immunohistological study of the origin of LHRH-containing nerve fibers of the rat hypothalamus. *Brain Res.* **103,** 597–602.

Sharpe, R. M., Fraser, H. M., Cooper, I., and Rommerts, F. F. G. (1982). The secretion, measurement, and function of a testicular LH-RH-like factor. *Ann. N.Y. Acad. Sci.* **383,** 272.

Snyder, GD, Capdevila, J., Chacos, N., Manna, S., Falck, J. R. (1983). Action of luteinizing hormone–releasing hormone: Involvement of novel arachidonic acid metabolites. *Proc. Natl. Acad. Sci. U.S.A.* **80,** 3504–3507.

Thorner, M. O., Edwards, C. R. W., Hanker, J. P., Abraham, G., and Besser, G. M. (1977). Prolactin and gonadotropin interaction in the male. *In* "The Testis in Normal and Infertile Men" (P. Troen and H. R. Nankin, eds.), pp. 351. Raven, New York.

Ungerstedt, U. (1971). Stereotaxic mapping of the monoamine pathways in the rat brain. *Acta Physiol. Scand. (Suppl.)* **367,** 82.

Vale, W., Rivier, J., Vaughn, J., McClintock, R., Corrigan, A., Woo, W., Karr, D., and Spiess, J.

(1986). Purification and characterization of an FSH releasing protein from procine ovarian follicular fluid. *Nature (London)* **321,** 776–779.

Valenca, M. M., Ching, M., and Negro-Vilar, A. (1986). Central inhibitory effects of LHRH analogs indicate an inhibitory role of LHRH to modulate its own secretion. *Int. Cong. Neuroendocrinol.,* 1st, San Francisco, California, p. 109 (abst.).

Vijayan, E., and McCann, S. M. (1978a). Re-evaluation of the role of catecholamines in control of gonadotropins and prolactin release. *Neuroendocrinology* **25,** 150.

Vijayan, E., and McCann, S. M. (1978b). The effect of systemic administration of dopamine and apomorphine on plasma LH and prolactin in conscious rats. *Neuroendocrinology* **25,** 221.

Vijayan, E., and McCann, S. M. (1980). Effect of blockade of dopaminergic receptors on acetylcholine (Ach)-induced alterations of plasma gonadotropin and prolactin (Prl) in conscious ovariectomized rats. *Brain Res. Bull.* **5,** 23.

Wise, P. M., DePaolo, L. V., Anderson, L. D., Channing, C. P., and Barraclough, C. A. (1979). Evidence that the pituitary gland is the site of inhibitory action of porcine follicular fluid upon FSH secretion in the rat. *In* "Workshops on Ovarian Follicular and Corpus Luteum Function" (C. P. Channing, J. M. March, and W. A. Sadler, eds.), pp. 437. Plenum, New York.

Witkin, J. W., Paden, C. M., and Silverman, A. (1982). The luteinizing hormone–releasing hormone (LHRH) systems in the rat brain. *Neuroendocrinology* **35,** 429–438.

Wuttke, W., Honma, K., and Hilgendorf, W. (1980). Neurotransmitter–neuropeptide interaction. *In* "Brain and Pituitary Peptides" (W. Wuttke, A. Weindl, K. H. Voight, and R. R. Dries, eds.), pp. 190–201. Karger, Basel.

Chapter 6

Fundamentals of Male Reproductive Toxicity Testing

JAMES C. LAMB IV*

Office of Pesticides and Toxic Substances
U.S. Environmental Protection Agency
Washington, D.C. 20460

I. General Approaches to Male Reproductive Toxicity Testing

A. Purposes of Different Study Designs

In 1965, studies in mice reported that chlordecone (Kepone) would cause infertility and testicular toxicity (Good *et al.,* 1965; Huber, 1965); in 1975, men exposed occupationally to large amounts of Kepone suffered various effects, including infertility (Taylor *et al.,* 1978). As early as 1961, data were published in the scientific literature indicating that 1,2-dibromo-3-chloropropane (DBCP) exposure decreased spermatogenesis in rats (Torkelson *et al.,* 1961); by 1977, occupational exposure to DBCP was associated with decreased spermatogenesis, low seminal sperm counts, and decreased fertility (Whorton *et al.,* 1977). These incidents are noteworthy, not because the

*This work represents the opinions of the author and does not necessarily represent official Agency policy, and it was not written as part of official duties.

persons made infertile by those chemicals were particularly numerous, but because both occurred some time after the publication of animal studies, showing that such effects could result from exposure to the chemicals.

Despite the availability of animal data, the investigation of human infertility in these cases was not triggered by animal data, but by human observation. A more recent case indicates, however, that animal data are less likely to be ignored. Ethlene glycol monoethyl ether (EGEE) was shown to decrease testicular weight and spermatogenesis in male mice (Nagano *et al.,* 1979); a study was then conducted in humans that demonstrated significant decreases in sperm count, without yet demonstrating clinical signs, such as decreased fertility (National Institute for Occupational Safety and Health, 1986). This points to the use of animal data to identify occupationally related effects on male reproduction.

The studies in this chapter help provide data that predict the potential for a chemical to adversely affect male reproductive function. The extrapolation of such effects to humans is termed "risk assessment." The decision of how to use the information, including social, economic, and public health considerations, is "risk management."

This chapter will focus on the toxicological evaluation of male fertility and reproduction, rather than on the individual organs or organ systems, which are discussed in other chapters.

From the perspective of a toxicity testing strategy, a toxicologist would normally begin with a very complete or comprehensive protocol that tests simultaneously as many organ systems as possible. The toxicologist then narrows the scope of the investigation and increases the specificity of the test until reaching the level of specificity required in each case. An example of a narrow test would be one that evaluates only sperm motility or only certain hormone levels; this leaves many other potential targets untested and is, by itself, too narrow to serve as a satisfactory testing screen. The scope of testing with narrow tests can be expanded by creating a battery of tests. This might include more sperm-function end points, additional hormones, hormone receptor levels, new organ-system evaluations, or fertility evaluation. Just such an approach was proposed by Galbraith *et al.* (1982) and tested by Amann and Berndtson (1986). A comprehensive test would theoretically challenge all potential targets at one time, and the efficient comprehensive test would test as many targets, but use as few end points as possible, such as evaluation of only fertility or reproductive function after chemical exposure. The latter approach is more orthodox, and the reader is referred to various reviews of such systems (Collins, 1978; Palmer, 1981; Lamb, 1985; Lamb and Chapin, 1985; Marks, 1985).

The comprehensive test system often uses broad endpoints to avoid the cases where a compound appears to have no effect but actually is a reproductive toxicant. In other words, a good reproductive toxicity screening system minimizes the proportion of false negatives. It should be clear, however, that there are certain trade-offs. The scientific trade-off is the loss of detailed information as a test becomes more and more general in scope or coverage. Also, a decrease in the proportion of false negatives is usually accomplished by increased sensitivity of the test and accompanied by an increase in the proportion of false positive tests. One very effective strategy is to start the evaluation with a comprehensive test and narrow the scope to more and more specific tests as key questions are answered. The choice of experimental approach should be made after careful consideration, rather than unwittingly or by accident.

It is not practical to provide a single protocol for testing male reproductive toxicity that would serve all research needs, unless cost and time were not important. Some regulatory agency protocols, however, seem to have been designed to serve all needs anyway. It is imperative that the investigator consider how the data will be used, before the protocol is finalized. The experimental approach for a laboratory that is performing general screening of dozens of compounds is very different from one investigating a chemical or drug with a very specific target organ effect. The experimental design should reflect the differences.

The very specific data provide important insights into the mechanism of action of a male reproductive toxicant. Once the mechanism of action has been identified, or at least explained to the point that the toxicologist can correlate the effects on reproductive function with certain features of the molecule, the investigator can help design chemicals that avoid unwanted effects on reproduction, while retaining the desired activity and intended use. This is a fundamental interaction of toxicology and pharmacology: to develop important and safer chemicals or drugs.

Data from other studies, in particular chemical disposition or metabolism studies, provide important leads into the usefulness of certain animal models for predicting human effects. Toxicants often cause effects only after being transported to a particular target organ or cell. The molecule may then be converted to a metabolite that causes the adverse effects. Therefore, it can be very important to know how the transport and metabolism of the chemical in the test animal compare to those in humans. Such data are often given significant weight by agencies performing risk estimations. These data help verify whether or not a particular strain or species is an appropriate model for human toxicity. Such data are highly chemical-specific and many demonstrate that rodents are superior to nonhuman primates for one test agent, but not another. Historically, such data do show that no one species other than humans has been universally better than all others for modeling human reproductive toxicity.

This chapter deals primarily with design of general reproductive toxicity test systems, the generation of safety data from such systems, and briefly discusses the use of the data in hazard identification and the eventual calculation of risk estimation. The data from the more general studies are often intended for risk assessment. Risk assessment is composed of hazard identification, dose–response characterization, exposure assessment, and risk characterization. It is important that studies used in risk assessment are reproducible and reliable models for human effects. The general studies are most often used, since they evaluate as many targets at a time as possible and since the general philsosphy is to overestimate, rather than underestimate, the potential of a compound to cause adverse effects based on data from animal models.

B. Type of Data Collected

The comprehensive laboratory tests for male reproductive toxicity are almost exclusively performed in rodent models; generally one uses either rats or mice. The number of animals required to generate useful statistical evaluations is too large for most other larger species, such as rabbits, to be used routinely. When larger species are used, the group size might be decreased for economic, rather than scientific, reasons, lowering the power of the test with a concomitant decrease in its usefulness. There is no statistical reason and generally no biological reason that the other species would be more reproducible or more sensitive to the chemical effects than rodents. Thus, there is no scientific basis for decreasing group size. Therefore, attempts at using other species may lead to inadequate tests, false negative results, or data that do not provide reliable margins of safety unless certain design characteristics are maintained, such as adequate sample size. The data collected on male reproductive toxicity include fertility (proportion of males which sire at least one litter), number of pups per litter, proportion of pups born alive, and survival and/or growth of the offspring to some specified age.

C. General Study Designs

Male reproductive toxicity is evaluated for risk assessment purposes in a manner that tests as many different target organs as possible in a single test. If there are significant adverse effects on fertility, they could be associated with effects on any of a number of target organs. For example, decrements in fertility may result from effects on testis function, endocrine toxicity, or accessory sex organ function or a combination of such effects.

The variables that are most fundamental in evaluating male reproductive function include route of exposure and dose levels, duration of chemical exposure before mating, fertility trial conditions, and endpoints monitored. These will be discussed below.

The route of exposure should generally be selected to model the human exposure, and it is usually the same route (oral, dermal, or inhalation) in the animal studies as the anticipated route for humans. Exposure to many chemicals would be expected to occure via multiple routes. For example, a pesticide, such as dibromochloropropane (DBCP), which has been used on food products, could result in occupational exposure to formulators and pesticide applicators through inhalation and/or skin absorption. For those eating food or drinking water containing residues, exposure is via the oral route. Animal studies have demonstrated that while a dose of DBCP given subcutaneously causes significant testicular toxicity, greater dose levels are required when it is given orally (Kluwe, 1981).

In some cases, the absorption, disposition, and metabolism of the chemical are critical in determining the toxicity of a compound. If the compound is rapidly metabolized and detoxified by the liver, then oral administration would not result in the same levels of active chemical at the liver and testis, since most of the blood leaving the gut goes directly to the liver. However, if the compound is absorbed through the skin or lungs, it will be distributed to the systemic organs before it is metabolized by the liver, and may result in a greater effective dose to the testis. In other cases, like ethylene glycol monomethyl ether (EGME), the route of administration makes little difference in the development of lesions in the testis. EGME is absorbed by the skin, lungs, or the gut, and while the metabolism of EGME occurs in the liver and other organs, it also occurs in the testis itself, and fertility is affected after exposure by various routes.

The timing of the chemical exposure compared to sampling is particularly important in male reproductive toxicity testing. As described in Chapters 2 and 7, the spermatogenic cycle is a process that takes about 8 weeks for development from spermatogonia to spermatozoa, plus about 2 weeks for epididymal transit. Any fertility assessment, by necessity, evaluates the function of ejaculated spermatozoa. If exposure to a test chemical affects the mature spermatozoa, then a mating trial conducted shortly after chemical exposure would be an appropriate test and fertility may be decreased immediately. By contrast, a chemical insult that causes an effect on the spermatogonia (early germ cells) would not be evident in a mating trial until those cells mature to spermatozoa and are ejaculated. Thus, effects on early germ-cell function may not be apparent as measurable effects on fertility until about 10 weeks after the initial exposure.

Many test designs deal with the time required for spermatogenesis by using an extended treatment period for the males followed immediately by a relatively short mating trial (Fig. 1). The extended exposure is intended to

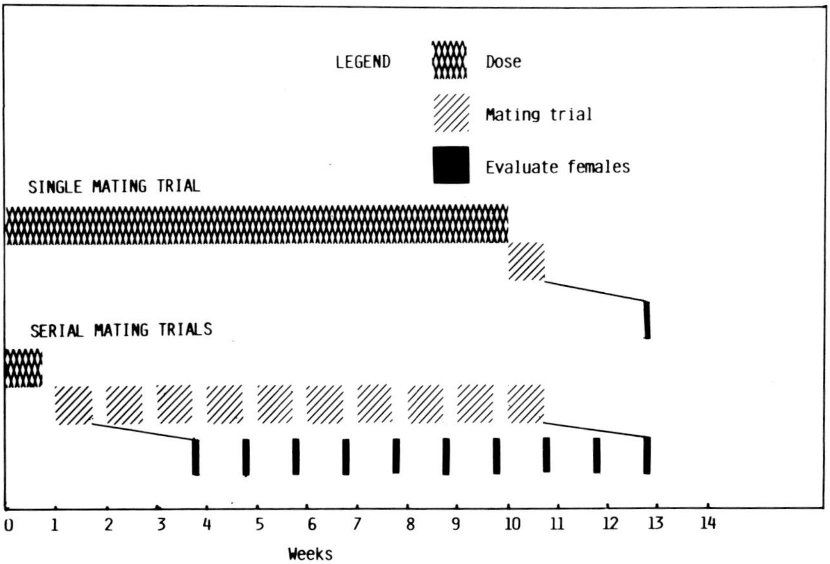

Fig. 1. Single-mating-trial and serial-mating-trial protocols. Single-mating-trial and serial-mating-trial protocols both require approximately 13 weeks to complete, but resource requirements are quite different. The single mating trial involves a 10-week dosing period for the males and then the one mating trial at the end of the dosing period. The serial mating trial has a short dosing period and then weekly mating trials for about 10 weeks after the treatment. In either protocol, the mating trial, here depicted as 5 days long, is followed by the evaluation of the female by necropsy late in gestation and counting the live and dead fetuses and corpora lutea.

ensure that the sperm ejaculated during the mating trial have been exposed to the test chemical through all stages of spermatogenesis. This approach is often used in regulatory agency protocols. It has the advantage of only requiring a single mating trial and conserves animals compared to a serial mating protocol (Fig. 1). The Food and Drug Administration (FDA), Environmental Protection Agency (EPA), and the Organization for Economic Cooperation and Development have all adopted certain variations of this strategy for different regulatory purposes.

Such designs have been used in the FDA Segment I studies for testing drugs, the FDA multigeneration study for food additives, the EPA two-generation study for pesticides (Fig. 2), and other fertility studies including prolonged dosing and a single mating trial. This is an efficient design and allows one to evaluate the full spectrum of stages of spermatogenesis through a single mating trial, rather than a series of mating trials to cover individually the many stages of spermatogenesis. A single mating trial does not, however, allow the investigator to determine which stage (or stages) of spermatogenesis is affected by the chemical exposure. This comprehensive approach is popular and appropriate for risk assessment purposes, since it tests as many sensitive stages of spermatogenesis as possible in a single mating trial. It is also designed to identify effects on the developing reproductive system by testing the second generation, which was exposed continuously from conception.

A different approach in the evaluation of effects on male fertility and reproduction is the serial mating trial (Fig. 1). Rather than a prolonged dosing followed by a single mating trial, the serial mating protocol uses a short-term treatment, usually 5 days, and a series of weekly mating trials. The mating trials are conducted each week for about 10 weeks. These studies provide important data on the sensitive stage of spermatogenesis by measuring the interval between the time of the exposure and the occurrence of significant changes in reproductive function. For example, an effect in the

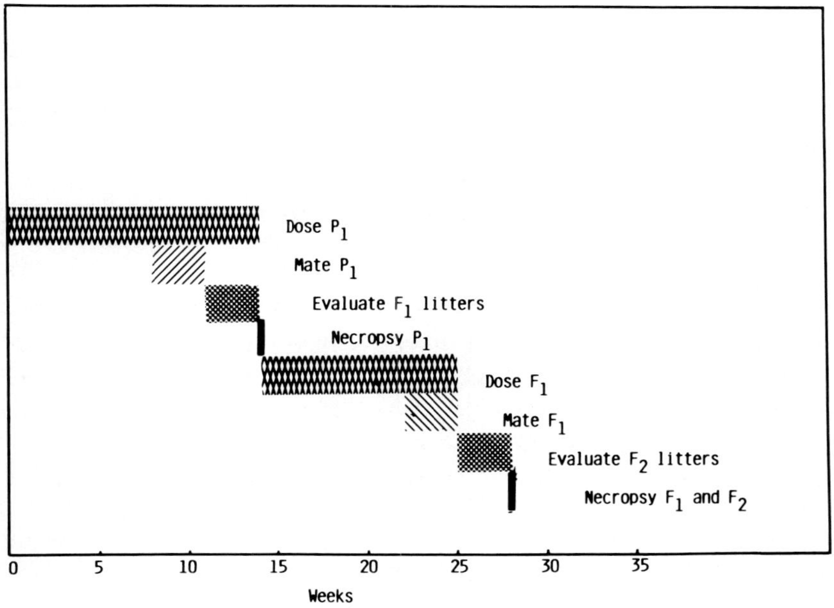

Fig. 2. EPA two-generation study. The standard two-generation protocols for pesticides and toxic substances start with a prolonged treatment of the animals and then a mating. The pregnant females (P1) are allowed to deliver and rear their litters. Members of the second generation (F1) are maintained on the same treatment regimen until sexual maturity and then are also mated, and the third generation (F2) is delivered and then maintained through lactation. There are necropsies scheduled for each of the generations, and the reproductive toxicity of the compound is tested by a single mating trial for each of two generations. The second generation has been exposed to the test compound through all stages of development.

first week would be related to changes in epididymal or vas deferens spermatozoa. Changes after 8 – 10 weeks would be associated with effects on spermatogonia that divide and mature in the subsequent weeks. The correlation between the spermatogenic cycle and the time until effects are observed is quite predictable within a given species and strain, since the timing of the spermatogenic cycle is fixed and proceeds at a certain rate, once the cells begin to mature into spermatozoa.

The serial mating protocol is quite successful at identifying the stage of spermatogenesis affected by the treatment, as long as the dose levels are properly selected. If the dosage levels are too low, the effect will not be detected by the weekly mating trials. If they are too high, the spectrum of affected germ-cell stages increases and the most sensitive stage cannot be precisely identified. For example, at a dose level of 100 mg EGME/kg day for 5 days, there were significant effects on the number of live pups per litter and fertile matings in only the fifth week after treatment. At 200 mg EGME/kg day for 5 days, however, there were significant effects at weeks 4, 5, 6, 7, 8, and 16 (Chapin *et al.,* 1985).

Both the prolonged dosing studies with the single mating trial and the serial mating studies have certain common features. In each case fertility is evaluated in a limited mating trial of 5 – 14 days, and each male sires no more than one litter per female. Although an individual male may be mated to more than one female in a given week (e.g., a group of 20 males may be mated to 40 females), the effects on fertility, live offspring per litter, fetal mortality, and other parameters measured in relation to the female should be evaluated statistically in relation to the male as the experimental unit. The number of females serves as the group size when they are treated and are the experimental unit. An error in the designation of the group size (e.g., when the male is dosed and mated to two females and the number of females is used as the group size) would lead to an overestimation of the statistical significance of the results (Weil, 1970). The advantage of mating one male to more than one female is not greater statistical power from an increase in the statistical group size, but is the increased reliability of the test (in the event that a female is sterile), or improved discrimination of effects by evaluating the ability of the male to inseminate more than one female in the mating trial and to sire more than one litter in that period.

The majority of the test protocols preferred by regulatory agencies use a prolonged dosing and a single mating trial wherein each male is housed with one or two females and may sire a couple of litters in that trial. Such mating trials are scheduled once or twice within the study, and the males are housed apart from the females for the rest of the study. The single mating trial is more popular since it uses significantly fewer female animals and an equal number of males. Both protocols use about 80 males in four groups (control and three treated groups) of 20, and the single mating trial requires 160 females if mated at two females per male. In consideration of resources, the serial mating requires 160 females each week of the 10-week serial mating, or 1600 female animals. Unless the test chemical and dosing are particularly expensive (some specialty chemicals can be quite expensive and inhalation exposures are costly), the added labor and chemical costs associated with the longer dosing will easily be recovered by using 1460 less females in the single mating trial. The serial mating and the use of the additional animals should be reserved for studies where determining the affected stage of spermatogenesis is particularly necessary, or when shortening chemical exposure time is advantageous. Furthermore, it should be noted that the affected stage may be determined by biochemical or morphological or other methods that are less expensive than serial mating trials (see Chapters 2 and 7).

Reproductive toxicity testing required by the FDA for new drugs includes several types of studies, but the protocols were drafted primarily to focus on problems like birth defects associated with thalidomide, and thus focus on developmental toxicology rather than fertility. These studies are referred to as the Segment I, II, and III studies, and only Segment I deals with male reproduction (Collins, 1978; Food and Drug Administration, 1966). Segment I is a rather straightforward variation on the single mating trial described in Fig. 1. The protocol calls for treating 10 males per group for 60–80 days (approximate time to cover all stages of spermatogenesis and epididymal transit) and then treating the 20 females per group for 14 days (minimum of two estrus cycles), then conducting the mating trial (one male to each two females). One-half of the females are killed at midgestation and uterine contents are examined for live and dead fetuses; the other half are allowed to deliver and the surviving pups are weighed and counted on days 1, 4, and 21. This uncomplicated design provides significant information on fertility, can be followed with more detailed studies, and is considered a screening system. The small numbers of males does cause the test to be less powerful than other designs (Schwetz *et al.*, 1980).

The EPA has adopted two sets of guidelines for two-generation study designs, but the protocols are very similar (Environmental Protection Agency, 1982a,b). The EPA and other agencies have changed their guidelines for such studies in the past, and the investigator is certainly advised to check the details at the time the study is going to be conducted, whether dealing with the EPA, FDA, or any other regulatory agency. The design of the current EPA protocol is given schematically in Fig. 2. It requires a dosing period of at least 8 weeks for mice and 14 weeks for rats followed by a mating trial for about 3 weeks. There are 20 males and at least 20 females for controls and three dose groups. The protocol calls for mating enough females to assure 20 pregnant females per group, but this can be logistically impossible if the compound is a reproductive toxicant. It is difficult to predict that the chemical may adversely affect fertility, and therefore it may not be possible to have 20 pregnant females in the test. In the EPA study, the second generation is held for at least 11 weeks for mice or 17 weeks for rats before a mating trial is conducted on the second generation. The design of the EPA protocol does evaluate fertility, development of the offspring, and organ toxicity for the first generation and then the reproductive development, fertility, and organ toxicity of the second generation.

The EPA design is a basic, no-frills, two-generation study requiring only one litter per generation and sticking to rather fundamental end points. The use of the second-generation fertility assessment, the organ toxicity data, and the larger number of males per group does significantly improve the evaluation of reproductive function, compared to the FDA Segment I study.

The FDA has different requirements for testing intentional food additives than for drugs. The multigeneration studies used in this case can be burdensome compared to the Segment I study for drugs and may require large resources and a number of years to complete. The protocol calls for three generations and one to three litters per generation (Collins, 1978). It is, however, not fundamentally different from any of the other multigeneration designs, and the mating trial is only conducted after the males have been treated for at least the equivalent of the time required for spermatogenesis. The second and third generation must be sexually mature before a mating trial can be conducted. As a result, the length of time required for such a study is substantial. The major controversy surrounding the three-generation study is whether the third generation really adds significant information, considering the cost and effort involved (Food and Drug Administration, 1970).

Another approach to evaluating fertility and reproduction in males uses a prolonged cohabitation of the males with females as breeding pairs and evaluating the cumulative fertility over a period of time. Such an approach, termed fertility assessment by continuous breeding (FACB), has been applied to mice and rats (Lamb, 1985; Lamb *et al.*, 1985). The protocol in mice calls for a 14-week period during which mating pairs (one male and one female) are cohabited continuously (Fig. 3). At birth, the offspring are removed, sexed, weighed (by litter and by sex, but not individually), and not returned to the cage, so that the female can immediately become pregnant again. At the end of the 14 weeks the males are separated from the females; most females will deliver one additional litter. That last litter can be used to evaluate a second generation.

Under such a scheme, each pair can deliver up to five litters in the time allowed. The average number of litters per pair gives a unique index of fertility or subfertility. The litter data from a given pair can be summed to

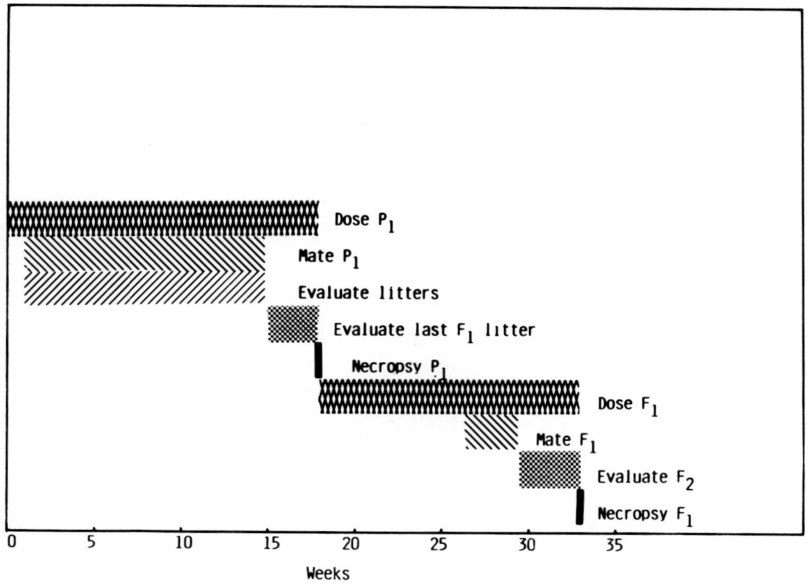

Fig. 3. Fertility assessment by continuous breeding (FACB). The FACB protocol shares most of the same time lines and endpoints with the EPA two-generation study. The major difference is that there is a greater effort toward collecting data on fertility of the first generation. The animals are housed as breeding pairs, and fertile matings may occur within a week of the first treatment. Unlike the other designs, the males and females are not separated; they are cohabited as breeding pairs through 14 weeks, and the offspring are removed within the first day after delivery and are evaluated. Each pair may deliver up to five litters each; the last litter may be delivered after the 14-week cohabitation period, and that litter may be kept until maturity and then reproductive performance evaluated in a single mating trial. Animals are necropsied from each generation. In addition to yielding data on the fertility or infertility of a given pairs, there is a more discriminating measure of fertility, the number of litters per pair.

give a better and more representative indication of the reproductive capacity of the test animals. Furthermore, the data are collected in such a way that fertility data are available on the effects of the chemical shortly after exposure, as well as 14 weeks after treatment. The latter time, 14 weeks, is more similar to the data presently collected in single mating trials following prolonged exposures.

Other studies of male reproductive function that are not at all routine are those on male-mediated teratogenesis—that is, the ability of a chemical exposure to the male to cause adverse developmental effects on the offspring that are sired by a subsequent mating. This approach has been attempted in relatively few cases and convincingly demonstrated in even fewer. Most recently, however, effects were demonstrated in the offspring of male mice exposed to cyclophosphamide (Trasler *et al.*, 1985). These studies typically involve the treatment of male mice or rats, mating those males to untreated females some time after the treatment, and either killing the females near the end of pregnancy to perform a teratological examination on the fetuses and identify birth defects, or allowing the female to deliver offspring and evaluating the morphological or behavioral features of the offspring.

D. Endpoints Evaluated

All of the comprehensive test systems described above generate similar information through different approaches. Those end points are described below, and they provide the data that are eventually used to make a risk assessment.

Mating is often evaluated in studies of male reproductive toxicology, and a mating index (number of copulations per number of estrus cycles required $\times$ 100; only one copulation is counted per estrus cycle) can be calculated. This serves as an indication of libido and is rather conveniently evaluated by checking for either vaginal plugs or sperm-positive vaginal smears. The data are often expressed simply as the proportion of males inseminating females, or the mean number of females inseminated per male if each male is housed with more than one female, or, perhaps incorrectly, the percent females inseminated. More elaborate evaluations may also be performed, as described in Chapter 8.

The fecundity index (number of pregnancies per copulations $\times$ 100) is also an index that is often used and indicates whether or not the males that mated actually impregnated the females. A reduction in this index may reflect changes in fertilization or implantation, and any implantation site would be sufficient to demonstrate that the animal was pregnant, whether or not the pregnancy resulted in live offspring. A male fertility index (number of males impregnating females per number of males exposed to fertile non-pregnant females $\times$ 100) is also used in many studies, and it does not rely on confirmation of mating.

The effects of chemical exposure on implantation and offspring development are also evaluated in many multigeneration studies. Some designs, like the standard dominant lethal design (Chapter 9), may call for sacrifice of the female before parturition, and the number of live fetuses, resorbed or dead implantation sites, and corpora lutea can be precisely counted. Counting corpora lutea gives an objective measure of the number of eggs ovulated by the female. It is useful indicator of the maximum number of offspring possible in a single mating, ignoring the possibility of twins. The number of corpora lutea can be compared to the total number of implantation sites; any decrease in implantation sites compared to corpora lutea is interpreted as a decrease in either fertilization or implantation of the fertilized ova. Increases in fetal death and resorbed implantation sites indicate postimplantation loss. As described in the discussion of studies of dominant lethal mutations in Chapter 9, postimplantation losses can result from, among other things, mutations of the male germ cell that are incompatible with life of the conceptus. Preimplantation losses can result from either mutations that interfere with the ability of the sperm to fertilize the egg or the process of implantation.

Multigeneration protocols generally include an evaluation of offspring survival, growth, and development. The EPA two-generation study, for example, requires study of the offspring at 1, 4, and 21 days of age. The data on survival can be expressed based on the number of pups born alive divided by the number of pups born. The day 1 survival divides the number alive on day 1 by the total number born. The day 4 survival divides the number alive on day 4 by the number born. The day 21 survival divides the number of pups still alive on day 21 by the number that were alive on day 4 (Harbison *et al.*, 1977).

Body weight of the growing offspring is another measure of toxicity. It should be noted, however, that body weight of the offspring can actually be increased by a toxicant rather than decreased. A common example of this is a toxicant that decreases the number of offspring per litter; the surviving offspring of smaller litters will be heavier than those from larger litters. This is because the surviving offspring do not have to compete for nourishment and grow at a faster rate than they would have in a normal-sized litter. This situation has little relevance to the human situation, but the investigator should be aware of this phenomenon so that pup weight changes are not misinterpreted. Also, certain toxicants may directly affect the pups, causing increases in pup weight, for example, by causing generalized edema, but generally one expects pup weight to be decreased by toxicant exposure.

Other endpoints have also been incorporated to study sexual or behavioral development. For example, testicular descent can be monitored in prepubertal males and vaginal opening in females. Both are markers of puberty. Also, certain motor-activity or general behavioral end points can be added to the study protocol. Such markers help relate reproductive toxicity to general toxicity.

Target organ effects are often included in the evaluation of male reproductive effects (Amann, 1982; Meistrich, 1982). The target organ may be identified and studied by a wide variety of approaches. Organ weight and morphology are among the most often used methods to decide whether or not the testes, or accessory sex organs, are affected. These methods are inexpensive and relatively rapid. They can generally be incorporated into standard multigeneration or subchronic toxicity studies for a small increase in cost. Sperm assessment may include sperm motility, count, and morphology; these techniques are discussed in some depth in Chapter 8. Once the target organ is identified, additional investigation may follow to further explain the toxic response. Many such studies include pharmacokinetic studies, which may help decide whether or not the animal model is relevant to predict human toxicity. Since many toxicity tests may be conducted at relatively high doses compared to expected human exposures, it can be important to determine whether or not the animal model converts the test agent to representative metabolites and that excretory mechanisms are not overloaded by the higher doses that may or may not occur in human cases. Knowledge of the biological processes in the animal model can help ensure that the end points that are measured are equivalent or are reasonable models of processes in humans.

II. Risk Assessment and Male Reproductive Toxicology

Once a chemical has been demonstrated to affect male reproductive function or target-organ changes in an animal model, it is important to determine whether or not the effects are relevant to humans and whether they predict potential adverse effects in humans. This prediction is the risk assessment. It does not predict with particular certainty the number of people who will be made infertile by the compound, although such data are encountered in the popular press, but it gives the risk manager a sense of the potential for the chemical to cause significant public health problems. It is one of many factors in managing the production and use of chemicals.

The risk assessment process involves a number of steps. Showing that the animal may be affected by chemical exposure is only the hazard identification step (National Academy of Science, 1983) and three other steps follow in the risk assessment. Once the risk assessment is completed, the risk management decisions remain, and those involve numerous social policy issues and are outside the scope of this chapter (see Nicholson, 1981; Department of Health and Human Services, 1986). After hazard identification, the risk assessment process is comprised of a dose–response assessment, exposure assessment, and risk characterization.

While hazard identification is the answer to a yes–no question (does the compound affect male reproduction?), the dose–response assessment is the evaluation of the amount of chemical exposure required to induce the identified response. It is important to determine the relationship between the dose and the measured response and factors or variables that may affect the response. In the case of a multigeneration study, the data are important in the hazard identification first and, if positive, the dose–response assessment second.

The response, as well as the dose, must be considered. For example, more weight or concern is given to irreversible infertility than to transient changes in sperm count. But such considerations are primarily risk management, rather than risk assessment, decisions.

The exposure assessment defines the exposure currently experienced by the human population and relates the conditions of the human exposure to the animal exposure. It must include more than amount or route of exposure. It must include also the duration and frequency of exposure and information on the population exposed. One major regulatory development that has recently been implemented to ensure that toxicity studies do provide adequate exposure data is good laboratory practice regulations (GLPs). GLPs require the confirmation that the chemical of interest is the one being tested, that the dose levels are actually being administered in the amount desired, and that protocols and standard operating procedures are developed and followed.

Risk characterization is the final step of the risk assessment process and is entirely dependent on the quality and integrity of the previous steps. Any error in the earlier steps will be carried over to this last step in the risk assessment. Risk characterization estimates the probable incidence that the adverse effects observed in the test animal will be present in the population being modeled. Its usefulness is limited to where there are reasonable human exposure data, and the failure to have adequate animal exposure data makes even the best toxicity tests unusable for proper risk assessment.

The entire risk assessment process is based on a series of educated assumptions. The risk management process that follows the risk assessment is based, in part, on the confidence that the risk manager has in the risk projections that come out of the risk assessment. Considering DBCP as an example of the process does illuminate certain problems that must be pointed out. In the case of DBCP, there were animal data that provided the hazard identification and the dose-response assessment. A major data gap, however, prior to the identification of human effects, was an exposure assessment for the human workers and a complete risk assessment. Since human effects were identified, DBCP exposures have been lowered substantially, but it is preferred to make the risk assessment before the risk management decision. The human exposure data are often the weak link in the risk assessment chain, followed only by a failure to use or believe or understand the hazard identification data.

By contrast, studies on human male reproductive endpoints were combined with EGEE exposure data and identified decreased sperm counts associated with exposure to EGEE at levels an order of magnitude below occupational standards (National Institute for Occupational Safety and Health, 1986). The human studies were conducted because of concerns raised by studies in laboratory animals (Nagano *et al.,* 1979). The NIOSH studies were then conducted on sensitive leading indicators (i.e., sperm count, sperm motility) that may be affected prior to observable effects on fertility. Such an approach is an excellent example of how male reproductive toxicity studies not only predict accurately reproductive hazard, but also lead to human studies demonstrating effects that can be presented by eliminating, or decreasing, exposure.

DBCP is particularly illustrative of two common phenomena. First is the failure to take notice of hazard identification data pointing to a problem. Second, the observation of reproductive risks is not limited to highly trained observers. It was a group of workers who deductively concluded that DBCP, or some other agent used on the job, was responsible for their low fertility. It is this type of observation, either by the victim or an astute observer, that has identified most human reproductive toxicants to date (Omenn, 1983). The other sources are laboratory studies as described or human epidemiological studies. All of these sources of information must be considered, together with relevant data on chemical use and exposure data, to determine the potential risk of the chemical to male reproductive function.

References

Amann, R. P. (1982). Use of animal models for detecting specific alterations in reproduction. *Fundam. Appl. Toxicol.* **2,** 13–26.

Amann, R. P., and Berndtson, W. E. (1986). Assessment of procedures for screening agents for effects on male reproduction: Effects of dibromochloropropane (DBCP) on the rat. *Fundam. Appl. Toxicol.* **7,** 244–255.

Chapin, R. E., Dutton, S. L., Ross, M. D., and Lamb, J. C. (1985). Effects of ethylene glycol monomethyl ether (EGME) on mating performance and epididymal sperm parameters in F344 rats. *Fundam. Appl. Toxicol.* **5,** 182–189.

Collins, T. F. X. (1978). Multigeneration reproduction studies. *In* "Handbook of Teratology" (J. G. Wilson and F. C. Fraser, eds.), Vol. 4, pp. 191–214. Plenum, New York.

Department of Health and Human Services (1986). "Determining Risks to Health: Federal Policy and Practice." Auburn House Publishing Co., Dover, Massachusetts.

Environmental Protection Agency (1982a). "Reproduction and Fertility Effects. Health Effect Test Guidelines." EPA publication 560/6-82-001.

Environmental Protection Agency (1982b). "Pesticide Assessment Guidelines. Subdivision F, Hazard Evaluation: Human and Domestic Animals." EPA Publication 540/9-82-025.

Food and Drug Administration (1966). "Guidelines for Reproduction Studies for Safety Evaluation of Drugs for Human Use."

Food and Drug Administration (1970). Advisory committee on protocols for safety evaluations: Panel on reproduction report on reproduction studies in the safety evaluation of food additives and pesticide residues. *Toxicol. Appl. Pharmacol.* **16,** 264–296.

Galbraith, W. M., Voytek, P., and Ryon, M. G. (1982). "Assessment of Risks to Human Reproduction and Development of the Human Conceptus from Exposure to Environmental Substances." EPA publication 600/9-82-001.

Good, E. E., Ware, G. W., and Miller, D. F. (1965). Effects of insecticides on reproduction in the laboratory mouse. Kepone. *J. Econ. Entomol.* **58,** 754–757.

Harbison, R. D., Khera, K. S., and Wright, P. L. (1977). Reproduction and teratogenicity tests. *In* "Principles and Procedures for Evaluating the Toxicity of Household Substances." National Academy Press. Washington, D.C.

Huber, J. J. (1965). Some physiological effects of the insecticide Kepone in the laboratory mouse. *Toxicol. Appl. Pharmacol.* **7,** 516–524.

Kluwe, W. M. (1981). Acute toxicity of 1,2-dibromo-3-chloropropane in the F344 male rat. I. Dose–response relationships and differences in routes of exposure. *Toxicol. Appl. Pharmacol.* **59,** 84–95.

Lamb, J. C. (1985). Reproductive toxicity testing: Evaluating and developing new testing systems. *J. Am. Coll. Toxicol.* **4,** 163–171.

Lamb, J. C., and Chapin, R. E. (1985). Experimental models of male reproductive toxicology. *In* "Endocrine Toxicology" (J. Thomas, K. Korach, and J. McLachlan, eds.), pp. 85–115. Raven, New York.

Lamb, J. C., Jameson, C. W., Choudhury, H., and Gulati, D. K. (1985). Fertility assessment by continuous breeding: Evaluation of diethylstilbestrol and a comparison of results from two laboratories. *J. Am. Coll. Toxicol.* **4,** 173–184.

Marks, T. A. (1985). Animal tests employed to assess the effects of drugs and chemicals on reproduction. *In* "Male Fertility and Its Regulation" (T. Lobl and E. Hafez, eds.), pp. 245–267. MTP Press Limited, Boston.

Meistrich, M. L. (1982). Quantitative correlation between testicular stem cell survival, sperm production, and fertility in the mouse after treatment with different cytotoxic agents. *J. Androl.* **3,** 58–68.

Nagano, K. L., Nakayama, E., Koyano, M., Oobayashi, H., Adachi, H., and Yamada, T. (1979). Mouse testicular atrophy induced by ethylene glycol monoalkyl ethers. *Jpn. J. Ind. Health* **21,** 29–35.

National Academy of Science (1983). "Risk Assessment in the Federal Government: Managing the Process." National Academy Press, Washington, D.C.

Nicholson, W. J., ed. (1981). "Management of the Assessed Risk for Carcinogens." New York Academy of Sciences, New York.

National Institute for Occupational Safety and Health (1986). "Health Hazard Evaluation Report, HETA 84-415-1688, Precision Castparts Corporation, Portland, Oregon."

Omenn, G. S. (1983). Environmental risk assessment: Relation to mutagenesis, teratogenesis, and reproductive effects. *J. Am. Coll. Toxicol.* **2,** 113–123.

Palmer, A. K. (1981). Regulatory requirements for reproductive toxicology: Theory and practice. *In* "Developmental Toxicology" (C. Kimmel and J. Buelke-Sam, eds.), pp. 259–287. Raven, New York.

Schwetz, B. A., Rao, K. S., and Park, C. N. (1980). Insensitivity of tests for reproductive problems. *Environ. Pathol. Toxicol.* **3,** 81–98.

Taylor, J. R., Selhorst, J. B., Houff, S. A., and Martinez, A. J. (1978). Chlordecone intoxication in man. I. Clinical observations. *Neurology* **28,** 626–630.

Torkelson, T. R., Sadek, S. E., and Rowe, V. K. (1961). Toxicologic investigations of 1,2-dibromo-3-chloropropane. *Toxicol. Appl. Pharmacol.* **3,** 545–549.

Trasler, J. M., Hales, B. F., and Robaire, B. (1985). Paternal cyclophosphamide treatment of rats causes fetal loss and malformations without affecting fertility. *Nature (London)* **316,** 144–146.

Weil, C. S. (1970). Selection of the valid number of sampling units and a consideration of their combination in toxicological studies involving reproduction, teratogenesis or carcinogenesis. *Food Cosmet. Toxicol.* **8,** 177–182.

Whorton, D., Krauss, R. M., Marshall, S., and Milby, T. H. (1977). Infertility in male pesticide workers. *Lancet* **2,** 1259–1261.

Chapter 7

Morphologic Evaluation of Seminiferous Epithelium of the Testis

ROBERT E. CHAPIN*

Developmental and Reproductive Toxicology Group
Systemic Toxicology Branch
National Institute of Environmental Health Sciences
Research Triangle Park, North Carolina 27709

Most chemical toxicity evaluations start with a general pathology study. Frequently at this early stage in the evaluation of an unknown compound's toxicity, little is known about the mechanism of action. Thus, what is needed is a general definition of the target areas in the body. While histology cannot reveal much about the mechanism of action of a compound, it can provide very useful information to direct subsequent studies specifically to affected cells or regions. This is certainly true in the testis. Although many screening systems are being developed and validated to replace whole animals in testing, the best general tool for detecting testicular toxicity is undoubtedly a good histopathologic evaluation.

This chapter will discuss the kind of information that can be gained from histologic studies and some considerations for their execution, and finally will review some examples of stage-specific and cell-specific lesions that have

*This work was not performed as part of official duties, and does not necessarily represent the opinion of the U.S. Government.

been described for the seminiferous epithelium. The discussion will focus on the testis, and will demonstrate how informed study design can reveal a great deal about where in this complex tissue a compound acts. It should become obvious to the reader as the chapter progresses that an understanding of the physiology and anatomy of the testis will greatly aid in defining toxic effects. The reader is referred to Chapter 2 in this volume for an overview, and to some deservedly classic reviews on the subject by Steinberger and Steinberger (1975) and Fawcett (1975), and reviews on the toxicologic aspects by Sever and Hessol (1985) and Waller *et al.* (1985).

I. Why Morphology?

One almost confounding feature of the testis is that, if there is severe toxicity for a period of time sufficient to induce germ-cell loss, the weight of the testis falls significantly, a decline that will continue until there are no more germ cells available to lose (Cater *et al.,* 1977). At such a time, one derives just as much information from organ weight as one does from a histologic evaluation (see, e.g., Hashimoto *et al.,* 1981). Not surprisingly, more than one paper has appeared that purports to examine testicular toxicity, while few endpoints are examined other than testicular weight (e.g., Curto and Thomas, 1982). As a first step in evaluating a new compound this may be sufficient, but any detailed studies demand histologic examination. There are two important additional points. First, when looking at a new compound, one runs the risk that there will be significant tissue damage with no change in weight, as has indeed been shown (Chapin *et al.,* 1984a). Second, an end-stage lesion tells one that the testis is damaged, but it ignores the prospect that any one of the many cell types may be the initial target for toxicant action. Just as "all roads lead to Rome," one can initially target just one of many different cell types, and still wind up with the end-stage lesion: an atrophied testis containing few or no germ cells. For example, Fig. 1 shows what an end-stage lesion can look like, containing only Sertoli cells and occasional early germ cells; this was obtained from a rat treated with 2,5-hexanedione. It is possible to produce a very similar picture in a rat treated with a toxic glycol ether at sufficient doses. The fact that a similar end-stage lesion can be produced by two compounds having very different initial effects and target cells (see Creasy and Foster, 1984; Chapin *et al.,* 1983) underscores the utility of appropriately timed and well-performed histologic investigations. These designs involve a single large dose or divided daily doses, coupled with serial sacrifice schedules subdividing the first 24 h, and thereafter on a daily basis. These studies can inform about the testis just as general histopathology studies inform about other organs: they pinpoint

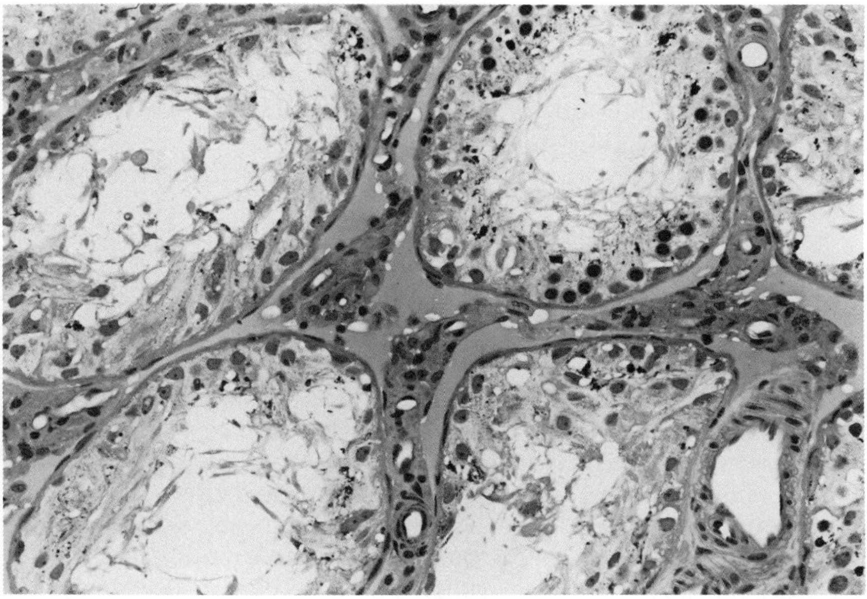

Fig. 1. Seminiferous tubules from a rat consuming 1% 2,5-hexanedione for 6 weeks. The tissue was fixed by perfusion *in situ* with aldehydes, embedded in glycol methacrylate, and stained with PAS and hematoxylin. While all four tubules contain one or more spermatogonia, only the one on the upper right contains any spermatocytes. × 350.

those cells that are sensitive to the administered compound by defining which of the many germ-cell stages or somatic cells in the testis appear to be initially affected. These effects can manifest as germ-cell death (Creasy and Foster, 1984), vacuolation of the Sertoli cell (e.g., Kierszenbaum and Mancini, 1973), or a disruption in the normal progression of spermatogenesis [see Chapter 2 for a brief description of the spermatogenic cycle, and Russell *et al.* (1983) for an excellent example].

In summary, testis weight after an extended period of dosing can demonstrate that the testis has been affected, but does not inform about where the primary lesion occurred.

At this point, the reader should be convinced that histologic examination is essential for an informed understanding of where a compound acts in the testis. While knowledge of the target cell is useful by itself, it is of even more benefit when one can then isolate that cell type and investigate possible mechanisms *in vitro.* Thus, histology is both the means and the end; while the initial studies are ends in themselves, they also indicate which cells or compartments in the testis should be pursued for further study. For example, if two compounds produce a similar lesion ultrastructurally, and the mechanism of action is known for one, there is a strong rationale for investigating a similar mechanism for the other (Russell *et al.,* 1981). Thus, appropriate histologic examination can be considered the foundation on which biochemical and mechanistic studies rest.

Now, what follows is a consideration of how to obtain tissue sections that are worth examining.

II. Considerations and Methods

There are two important considerations when planning a morphologic evaluation of the testis. First, one must fix the tissue to preserve the structural detail, for study of structure is the whole object of histology. Next, the testis must be embedded in a medium that will allow for two things: (1) the section must be thin enough to provide optimal resolution and minimal overlap of cells and structures, and (2) the medium must be permeable to the stains that will be used. Now, the reader can make a choice of methods, depending on the final product desired. Some methods are listed below in ascending order of preservation and resolution. All of the immersion fixation procedures require pricking the capsule of the testis 15–20 times to increase the penetration of the fixative. Embedding in paraffin follows standard procedures, and the glycol methacrylate method follows the manufacturer's protocol. Since the relative merits of each have been discussed at length elsewhere (Lamb and Chapin, 1985; Chapin *et al.,* 1984b; Russell, 1983), and because detailed methodology is already available in the primary literature and textbooks, the following discussion will be limited to summarization.

A. Formalin–Paraffin

This procedure is illustrated in Fig. 2(a). Immersion fixation in 10% neutral buffered formalin, followed by paraffin embedding, is clearly a very poor way to preserve and embed a testis, due to the enormous shrinkage artifact that this procedure introduces. Both nuclear and cytoplasmic detail are obscured; this method is useful only for counting cells, and, as we have seen above, cell count correlates closely with the weight of the tissue.

B. Helly's–Paraffin

This procedure is illustrated in Fig. 2(b). If the media must be paraffin, Helly's produces somewhat fewer artifacts in the epithelium than does Bouin's, although total tubular shrinkage is slightly greater. The primary disadvantage of Helly's fixative is that it decomposes when stored longer

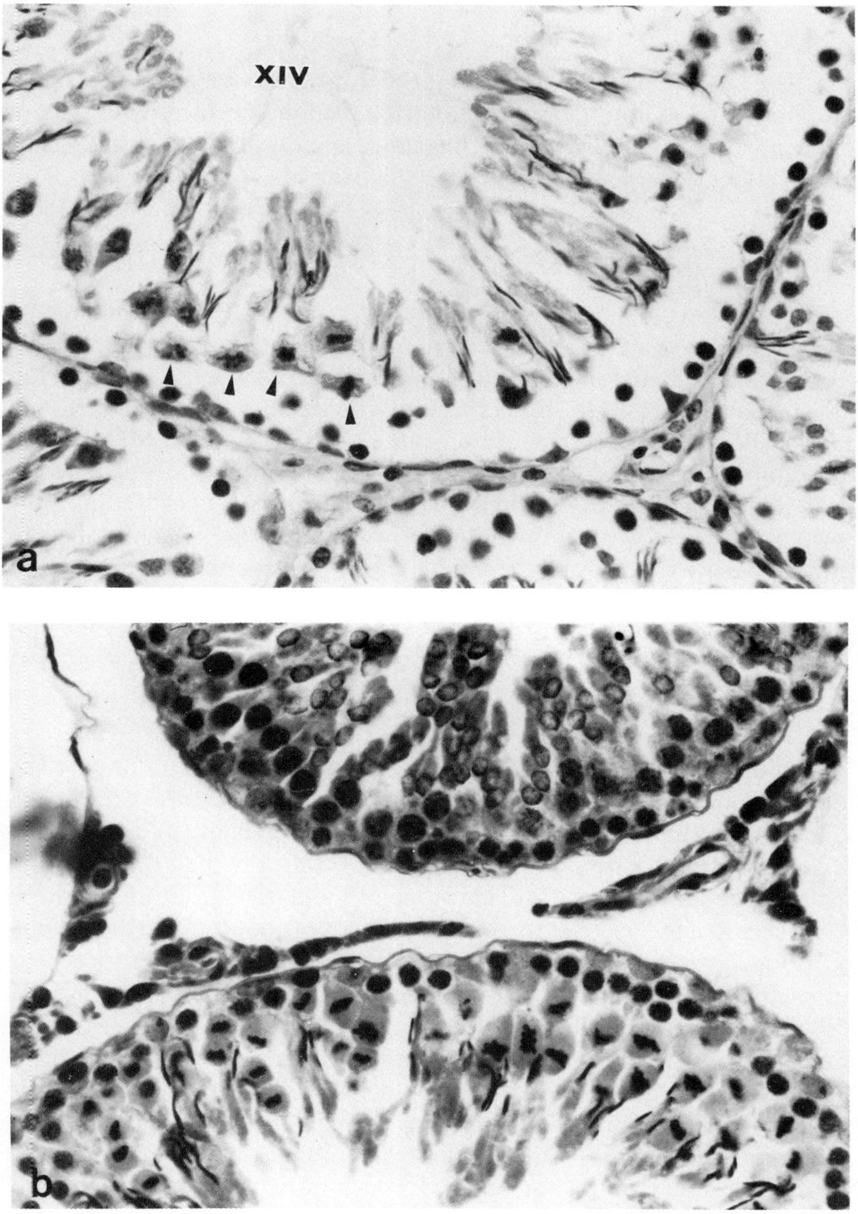

Fig. 2. (a) Seminiferous tubules from a testis immersion-fixed in neutral buffered formalin (NBF), embedded in paraffin. In the stage XIV tubule (XIV), meiotic spermatocytes in reductive division can be seen (arrowheads). Note the artifactual large spaces between the cells, and between the germ cells and the periphery of the tubule (where the arrowheads lie). × 700. (b) Stage IX (top) and stage XIV tubules from a testis fixed in Helly's fixative and embedded in paraffin. The germ-cell cytoplasm is less shrunken than when fixed in NBF (there are smaller spaces between adjacent cells), but tubular shrinkage is greater (larger spaces between adjacent tubules). There is very little "grain" in the cytoplasm. × 700.

than 36 h, so must be made fresh. It is probably this reason that drives many laboratories to use Bouin's as the immersion fixative of choice: a marriage of convenience that sacrifices some histologic quality for a decrease in the trouble taken to prepare and maintain the fixative.

C. Formalin–Glycol Methacrylate

The next step up in quality is immersion fixation in 10% neutral buffered formalin (NBF) as above and embedment into glycol methacrylate (GMA), a popular water-soluble plastic [Fig. 3(a)]. One can easily get sections of 2 μm with standard steel knives; thinner sections can be obtained with glass Ralph knives. The *bad* news is that, although the shrinkage is markedly reduced when embedding into GMA, it is not altogether eliminated. If one has the capability to make the transition from paraffin to plastic, I believe this is the most cost-effective way to obtain better histology on the large numbers of animals that are generally run through bioassay or large-scale studies, where vascular perfusion is impractical. The chemistry of formaldehyde fixation has recently been reviewed (Fox *et al.,* 1985).

D. Perfusion Fixation–Glycol Methacrylate

This step [Fig. 3(b)] marks the transition from the constraints imposed by large-scale studies to a more fine-tuned analysis of the structure of the organ. Generally, mixtures of formaldehyde and glutaraldehyde are perfused. A prewash of Ringer's balanced salts with 0.1% procaine and 0.1% sodium nitroprusside decreases the vascular tone of the testicular arteries and improves access of the fixatives to the organ. This is introduced into the ascending aorta, and leaves the body through an incision in the right ventricle. The fixative that follows is 4%(w/v) paraformaldehyde, 5% glutaraldehyde in 0.1 *M* sodium cacodylate to fix the tissues. A case could be made for performing whole-animal perfusion routinely, so that the preservation of all organs is improved. The disadvantages of this would be that (1) it is just too expensive in time and material to perfuse that many animals, and (2) because the testis suffers more than do most other organs from immersion fixation, the histology of other organs would not benefit as much from perfusion as would the testis. As an alternative to whole-body fixation, there are several methods for perfusing just the lower half of the animal, or cannulating the testicular vessels alone, or even performing subcapsular perfusion. The GMA is also permeable to water-soluble stains (unlike epoxy; see below), and allows histochemical procedures. Whole cross sections of testis can be embedded in GMA, a point of some value when counting the numbers of tubules in the various stages of spermatogenesis.

E. Perfusion – Epoxy Resin

This has been the subject of an excellent review by Russell (1983), to which the reader is referred. Aldehyde fixatives through the vasculature, and the tissue is embedded in an epoxy plastic such as Epon or Spurr's resin [Fig. 3(c)]. This represents the ultimate commitment to histologic preservation and resolution available with current techniques. The main benefits are those of truly breathtaking resolution and clarity, and the fact that one can examine a section with both the light microscope (LM) and electron microscope (EM). The drawback is that only a few tubules may be embedded in each block, which markedly increases the work needed to examine a whole testis. Furthermore, most of the stains commonly used to stain testis sections for the light microscope will not stain epoxy-embedded tissue, and most enzymes are destroyed by the embedding process. Nonetheless, the quality of the preservation is exceptional.

Many investigators believe that the earliest morphologic changes in the testis are found only by examination at the EM level. This has been shown to be true for the phthalates (Creasy *et al.*, 1983), where pathology was observed in the Sertoli cells 3 h after dosing, and the germ cells were still unaffected. There were no observable changes at the LM level at this time. It seems likely that our understanding of testicular histopathology would be greatly improved if studies designed for determining the affected cell type scheduled the first several time points at hourly, as well as daily, intervals. It is probable that the spectrum of effects is different at these early times from what is manifested 24 h after dosing.

The foregoing discussion has made it clear that there are a number of available strategies for obtaining good sections of testis; the direction taken depends on the balance between optimal structural preservation and other considerations (e.g., time and cost). For research applications, the appropriate technique would be vascular perfusion, followed by GMA for the LM, or epoxy resins for EM evaluation. If study size constraints limit the practicality of perfusion (as in larger studies aimed at general toxicity evaluations), the very least that should be done in Bouin's–paraffin; formalin–GMA is clearly an improvement on this and is no more difficult. It should be pointed out that a significant amount of what we know about the testis has come from studies using Bouin's and paraffin.

An alternative approach uses perfused tissue from a dose and time study, which one can embed first in paraffin for a general survey of effects. A preliminary examination of the paraffin-embedded tissues will then reveal which tissues should be processed into GMA for better resolution and finer examination. Sections from the same tissues can then be put into the epoxy media for examination under the electron microscope. Thus, the additional labor involved in perfusing the animals at sacrifice yields greater flexibility in

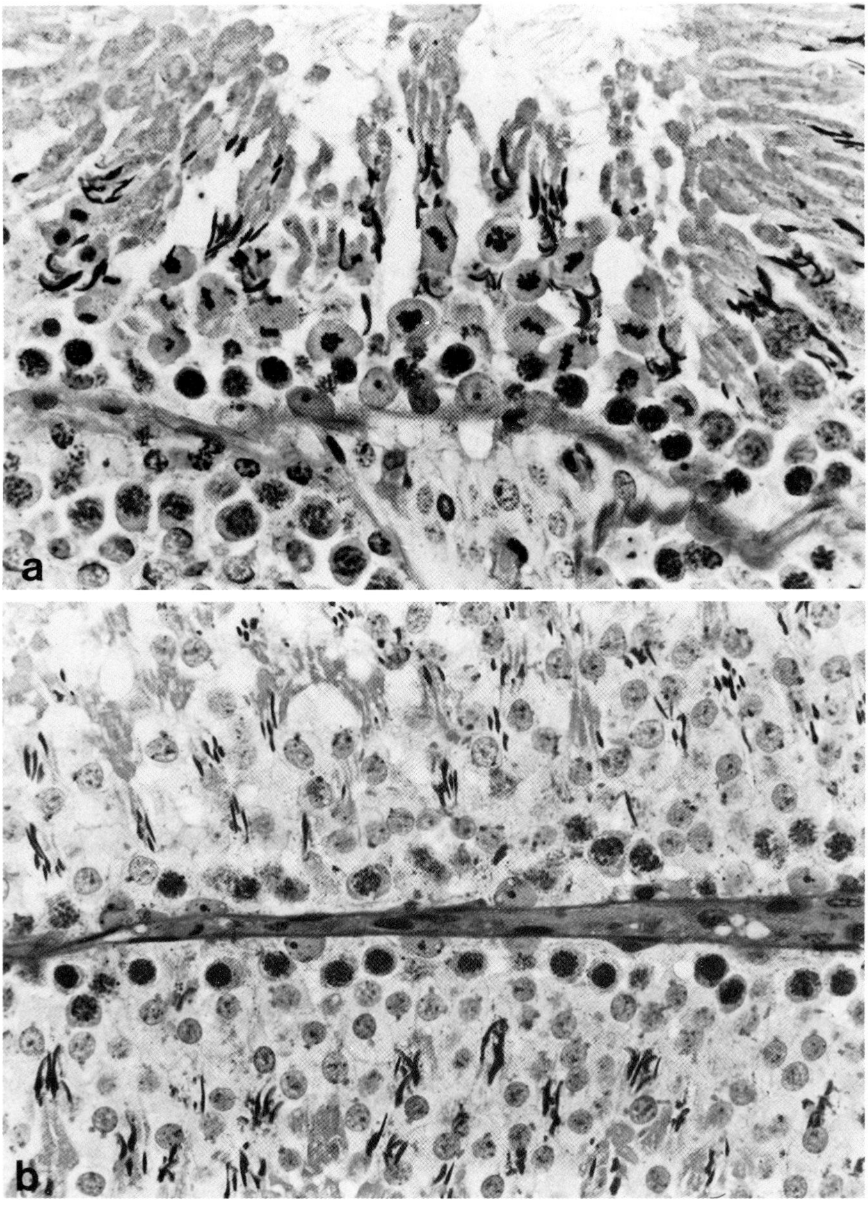

Fig. 3. (a) Stage XIV tubule from NBF-fixed testis embedded in GMA. Note moderate cellular shrinkage and lack of tubular shrinkage and cytoplasmic "grain." × 700. (b) Stage II (bottom) and III (top) tubules from aldehyde-perfused, GMA-embedded testis. The fixation prevents shrinkage; the thin (2 μm) GMA section permits greater resolution. × 700. (c) Stage II (top) and early VII (bottom right) tubules from aldehyde-perfused, epoxy-embedded testis. The wealth of subcellular detail is impressive. × 700.

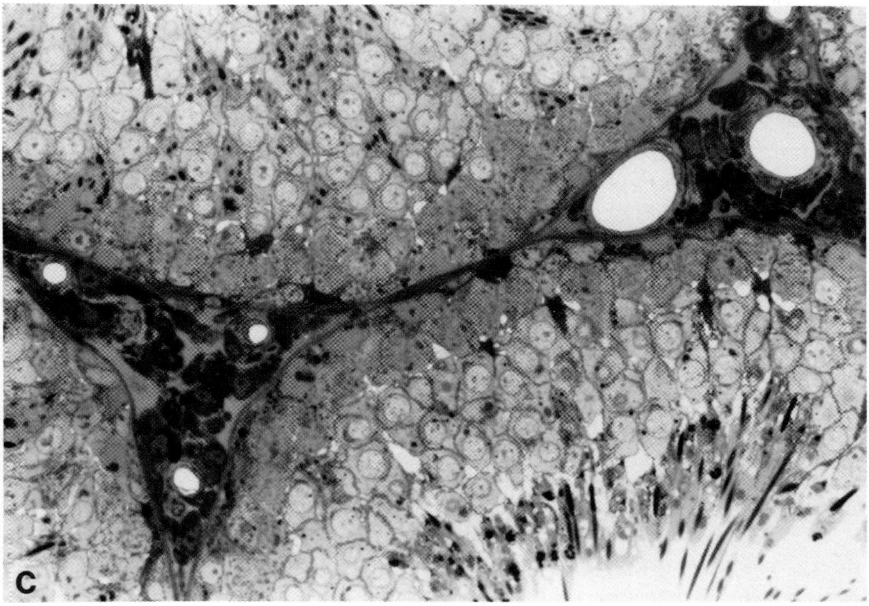

Fig. 3 (*Continued*)

subsequent processing of the samples and results ultimately in saving time; one need not dose additional animals, but can examine tissue from the same animals at different levels.

It should be pointed out that choice of stain is critical for proper evaluation of testis sections at the LM level. The stage of spermatogenesis of a given tubule is based, in part, on the changes in the acrosome and nucleus of the spermatids in that tubule (see Chapter 2; also see Leblond and Clermont, 1952). Eosin does not stain the acrosome of the developing spermatids. Thus, for the purposes of assigning lesions in the testis to particular stages, it is imperative that periodic acid–Schiff's reagent (PAS), followed by hematoxylin, be used. This regimen is no more difficult or time-consuming than that for hematoxylin and eosin, and should become standard procedure for light microscope–level evaluation of testis.

III. Media Effects and Glycol Methacrylate

The versatility of GMA as an embedding medium makes it a unique tool of the histologist. The following discussion will focus on (1) how the appearance of a germ-cell death lesion (necrosis) changes depending on the embedding media used and (2) a few of the special techniques that can be performed in GMA-embedded tissues.

The choice of embedding medium can determine how chemical-induced lesions will appear in the section. One possible concern about the immersion fixatives like Helly's and Bouin's is that they produce a variable amount of protein coagulation, which manifests as a "grain" in the cells (Zenker's is particularly bad in this respect, regardless of embedding medium). In GMA-embedded tissues, necrosis manifests as an increase in this visual "grain." Thus, it is possible that a routine artifact in the immersion-fixed tissues would interfere with a proper interpretation of a lesion. We wanted to determine whether the choice of embedding medium affected our ability to detect cell necrosis. We reasoned that delayed/inhibited release of late-stage spermatids would be relatively easy to detect even in severely shrunken sections of testis, while a lesion distinguished by changes in the appearance of the affected cells might be more difficult to see in markedly shrunken cells. One compound that produces predictable and defined germ-cell necrosis in the rat is ethylene glycol monomethyl ether (EGME). Four animals were treated with 0, 150, or 200 mg EGME/kg p.o., and sacrificed 24 h later. One testis from each rat was immersed into Bouin's, and the other into 10% NBF after the tunica was punctured 15–20 times with a needle. After a 24-h fixation period, the testes were sectioned, and cross sections of each organ were embedded in paraffin and GMA.

We found that the same lesion showed up in different ways in the different preparations. When tissue from either fixative was embedded in paraffin, we saw a change in the staining characteristics of the pachytene cytoplasm, turning from predominantly hematoxylin-positive to more PAS-positive (Fig. 4). This was easier to see in the Bouin's fixed tissue than the NBF-fixed tissue, partly because there was simply more cytoplasm available to stain. This change in staining pattern was used as one criterion in one of the original papers defining the testicular lesion of EGME (Creasy and Foster, 1984). In the tissue embedded into GMA, the intensity of the cytoplasmic staining changed, and the "grain" of both the cytoplasm and nucleus of the damaged cells was increased; the uniform staining normally found in the cytoplasm was replaced by an uneven, patchy stain distribution (Fig. 5). This was also seen previously (Chapin *et al.,* 1984c). Thus, the manifestation of the lesion was different in paraffin and GMA. The increased resolution in the GMA sections is apparent in these figures. These studies showed that the "grain" artifact did not interfere with detecting the lesion; the appearance of the lesion changed sufficiently to obviate this artifact. It should be noted that it is still possible that there was an initial increase in this "grain" in the paraffin-embedded sections, which occurred before the staining changes and was masked by the artifact in the tissue. Because the apparent extent of cell death was similar in sections of one testis embedded in the two media, this would seem unlikely.

Histochemical studies of the testis can be useful in identifying and localiz-

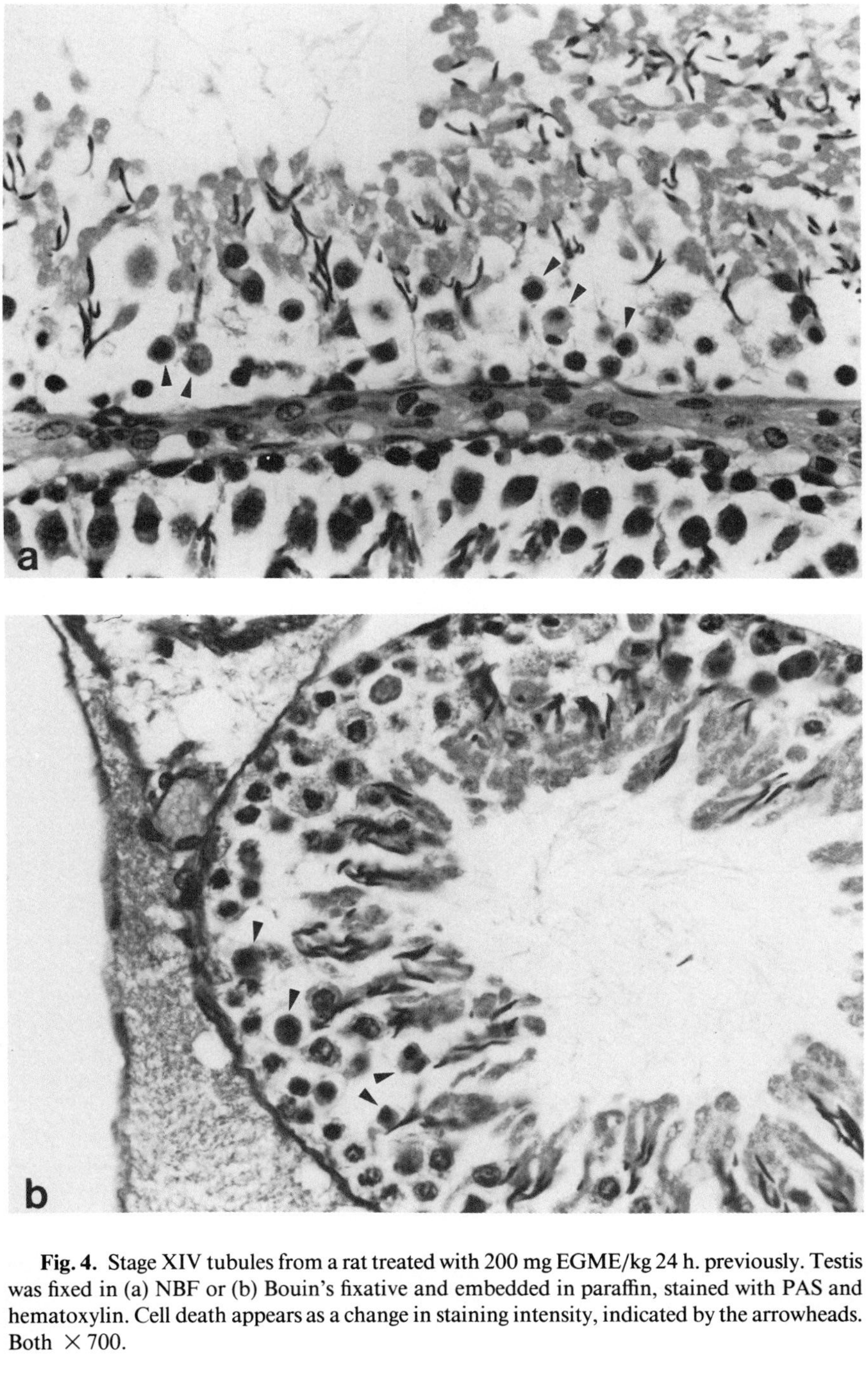

Fig. 4. Stage XIV tubules from a rat treated with 200 mg EGME/kg 24 h. previously. Testis was fixed in (a) NBF or (b) Bouin's fixative and embedded in paraffin, stained with PAS and hematoxylin. Cell death appears as a change in staining intensity, indicated by the arrowheads. Both $\times$ 700.

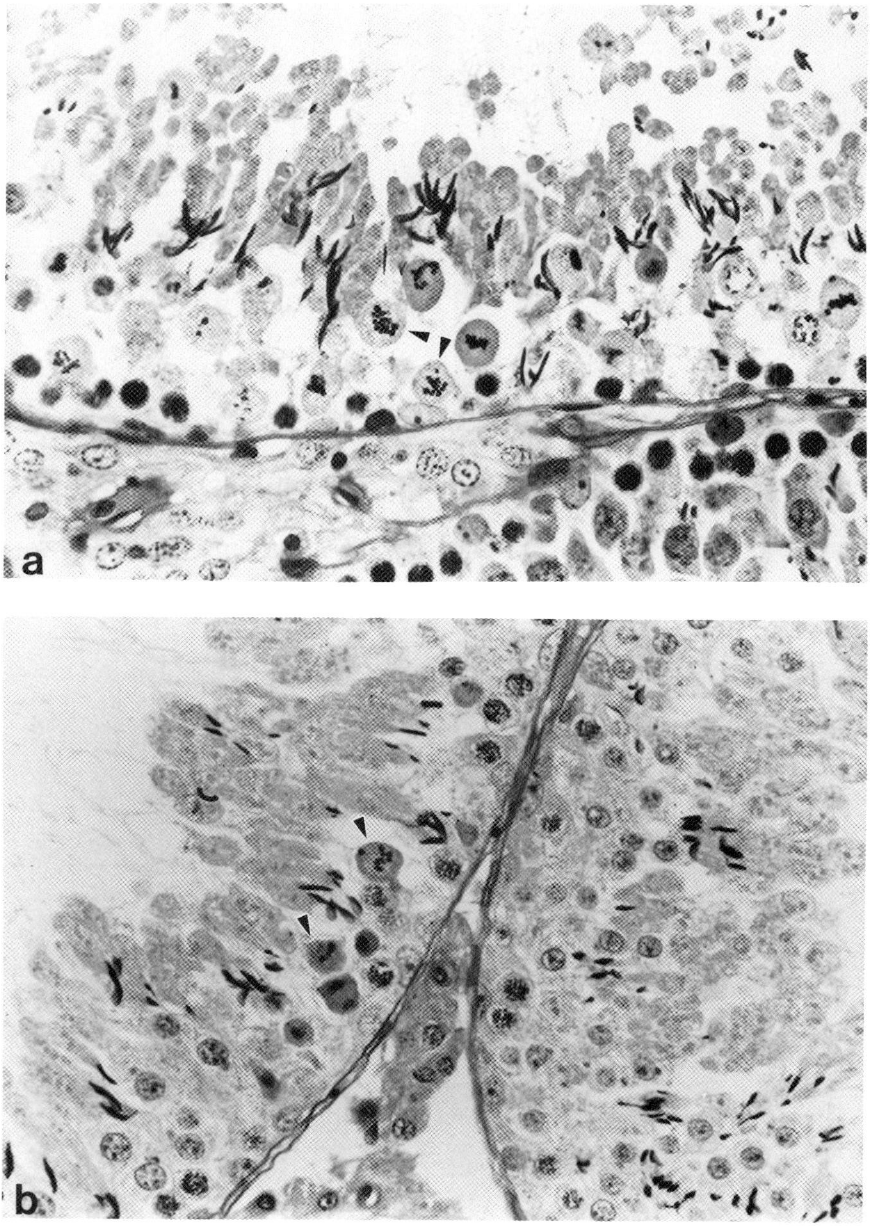

Fig. 5. Stage XIV tubules of testis from a rat treated as in Fig. 4, embedded in GMA. (a) In the NBF-fixed tissue, the dying cells are marked by the increased "grain" and fainter staining (arrowheads), while (b) in Bouin's-fixed tissue, necrotic cells stand out as more deeply stained (arrowheads). Both × 700.

ing enzyme activities. Even a casual glance at a testis section reveals that it is composed of many cell types, in tubular and extratubular compartments. Not surprisingly, there is a certain differential distribution of enzyme activities. Most frequently, this does not take an all-or-nothing form; rather, a large majority of total testicular activity of, say, nonspecific esterase will be localized to the interstitium (Kuehn-Velten *et al.,* 1982), while a much smaller amount is found in the tubules (Cameron and Snydle, 1985). When such enzymes are affected by toxicant exposure, histochemical examination of tissue from treated animals can be of use in localizing a primary site of toxicant action. Such data have been obtained in studies on the effects of tri-*o*-cresyl phosphate (TOCP), where histochemical studies on testes of TOCP-treated animals revealed that the major site of action was on interstitial nonspecific esterase (Fig. 6) (Somkuti *et al.,* 1986). These data are qualitative rather than quantitative, and serve to direct future studies. Similar kinds of data can be obtained for other membrane-bound enzymes found in a limited number of cell types. The majority of alkaline phosphatase (AP) activity is found in peritubular and endothelial cells (Kormano and Hovatta, 1972; Breitenecker *et al.,* 1974; Redi *et al.,* 1983), and reports of a decrease or increase in AP activity in homogenates (Scott and Persaud, 1977) suggest that qualitative histochemical studies might be of value here, also. These enzymes are well preserved in tissue that is fixed, rinsed, and embedded in GMA at temperatures not exceeding 4–5°C. Some cytoplasmic enzyme activities have been demonstrated in nasal epithelium processed at 4°C (Bogdanffy *et al.,* 1985). Details for performing such demonstrations are readily available (Bogdanffy *et al.,* 1986; Brinn and Pickett, 1979; Namba *et al.,* 1983).

Immunohistochemical studies are done with apparent ease in cold-GMA-embedded tissue (Pedraza *et al.,* 1984), although the testis has yet to really benefit from the full and creative use of such techniques.

One disadvantage of GMA is the tendency of the plastic matrix itself to stain with some basic dyes at neutral pH. Although not a problem for PAS or other routine stains, it can interfere with proper demonstration of di- and trichromic stains; a technique has been developed to remove this bothersome artifact (Hards, 1983).

IV. Examples of Stage-Specific Lesions

The discussion so far has addressed the possible ways to fix and stain testes for histology, and described briefly what each method produces. The testis is composed of many different cell types, and it is quite possible to produce selective lesions in just one or a few of these cells. This section will briefly

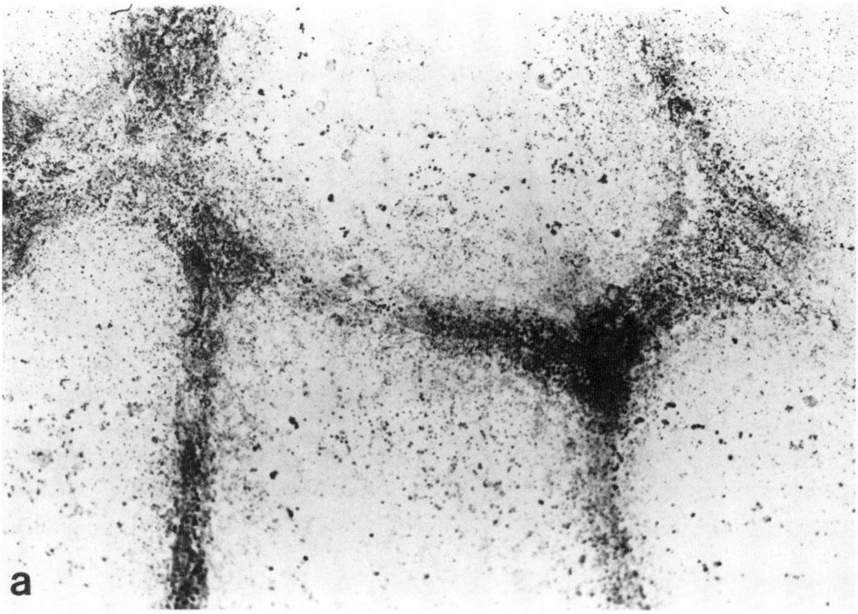

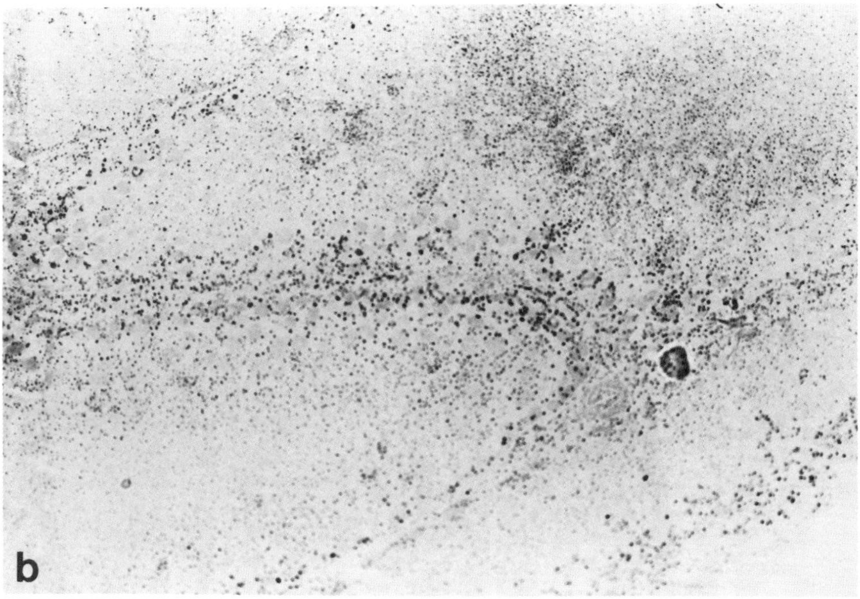

Fig. 6. (a) Tubules from control rat testis, frozen section, reacted for nonspecific esterase. Notice that the dark deposits indicating enzyme activity are concentrated mostly between the tubules. (b) Seminiferous tubules from rat treated for 3 days with 150 mg TOCP/kg. Notice that the intensity of the stain deposits has diminished markedly. Not counterstained. Both × 350.

describe some examples of studies that have identified the target cells in the testis for some selected treatments.

Generally, these studies at the LM level do not identify effects on Leydig cells, except insofar as the epithelium responds to a change in the local androgen status *(vide infra)*. Leydig cell changes that are not visible at the LM level have been recognized in the EM (Lee and Gillies, 1984). Morphometric EM studies done on epoxy-embedded tissue have shown that the volume of the Leydig smooth endoplasmic reticulum correlates with the androgen status of the animal (Ewing *et al.*, 1981; Mazzocchi *et al.*, 1982), and the volume of the Leydig cells varies with the stage of the adjacent tubules (Bergh, 1982); one could argue that these end points are more time-consuming and tedious than simply measuring testosterone, and yield essentially the same information. The message here is that LM-level effects on Leydig cells, other than obvious cell death, are difficult to see.

The general strategy employed to identify specific stages of spermatogenesis affected by toxicant exposure is to dose the animal for as short a period of time as practical, and then serially sacrifice the dosed animals at known intervals. Because it is known how long each stage of germ-cell development lasts (Clermont *et al.*, 1959), one can count cells at each maturational stage, count backward by the number of hours between treatment and sacrifice, and obtain a fair idea of which cells were affected by the treatment. This is one key, along with the proper materials, to good histologic examination of the testis: knowing how the stages of germ-cell development relate to each other temporally. The first two studies below exemplify this approach.

A classic example of a study that identified a select target-cell population is that of Chowdhury and Steinberger (1964), who described the lesion produced by heat. These authors used Bouin's–paraffin preparation, careful timing, and painstaking analysis to identify leptotene, zygotene, and pachytene spermatocytes, and the first two spermatid steps, as sensitive to the effects of 15 min of immersion into 43°C water. These authors also note that the spermatocytes and spermatids differ in the sequence of necrotic changes they undergo, differences that have been repeatedly observed by later authors. Spermatids formed ring-like nuclei, while the first visible change in spermatocytes was an increase in cytoplasmic PAS staining. The photomicrographs in this careful paper are characteristic of Bouin's–paraffin preparation. The design and execution of these studies can and should be applied to the analysis of more chemically induced lesions.

Another example is the lesion produced by ischemia (Steinberger and Tjioe, 1969). While interruption of blood flow for more than 2 h resulted in widespread cell death, ischemia lasting 90–105 min selectively affected type A spermatogonia in stage XII and XIV tubules, intermediate spermatogonia in stages I–IV, and spermatocytes in VII and VIII. There was maturation–depletion of germ cells, wherein the number of germ cells at subsequent stages was depleted by the loss of the precursor cells. A similar type of maturation–depletion effect is produced by EGME (Creasy and Foster, 1984) acting, as we have seen, on spermatocytes.

One study that described, without quantitation, some of the cells affected by chemical treatment is that reported by Hagenas *et al.* (1978), which used Bouin's–paraffin to identify the spectrum of effects produced by nitrofurazone. Other studies have used perfusion fixation and epoxy embedding to identify, without quantitation, the cells that are responsive to hypophysectomy (Russell and Clermont, 1977) or disturbance of the normal hormonal milieu of the testis (Russell *et al.,* 1981). Frequently, quantitation will help demonstrate the extent of the lesion, and will allow the reader to form an independent assessment of the severity of the effects. This omission notwithstanding, the fundamental utility of the studies by Russell *et al.* (1981) characterizing the appearance of the hormonally mediated lesion is that they form a yardstick by which the hormonal component of chemically induced lesions might be determined. In practice, this means looking for the range of affected cells in tissue from treated animals, and if these are wholly or partially similar to those affected by antiandrogens, it would suggest that compromised androgen status may play a part in the etiology of the epithelial lesion. Thus, epithelial changes can be secondary to primary effects in the Leydig cell, or higher levels in the androgen control pathway (pituitary or hypothalamus) (e.g., methyl chloride; Chapin *et al.,* 1984d).

One study in which counting cross sections of tubules at particular stages provided information not obtained by cell death or other abnormal morphology was performed with 2,5-hexanedione (Chapin *et al.,* 1983). This study found that there was a slowing in the progression of tubules through stages V and VI to VII, and showed how data can be obtained that are preliminary indicators of effects on the kinetics of spermatogenesis. The more straightforward morphologic evaluations of these testes revealed that the enlarged smooth endoplasmic reticulum, which manifested at the light microscope level as "vacuoles," was seen more frequently in some stages of spermatogenesis than in others.

One compound that produces death of specific cells is EGME; this lesion has been described in both Bouin's–paraffin tissues (Creasy and Foster, 1984) and perfused-GMA embedded testes (Chapin *et al.,* 1984c). In these two studies, the lesion appeared in the same cell types, but the nature of the lesion differed, as seen above, due to the different methods of preparing the tissue. Morphologically, EGME affects spermatocytes, both late pachytene and early spermatocytes. With slightly increased doses, and using fertility endpoints, one can determine that spermatogonia are also affected by EGME (Chapin *et al.,* 1985a).

These latter studies with EGME exemplify one property of spermatogonia; toxic effects are difficult to detect by morphologic criteria, and a frequent recourse is to measure the products of continued stem-cell division (number of germ cells in later stages). These other approaches to identifying spermatogonial toxicity are demonstrated in the studies investigating the resistance of stem cells to the cytotoxic effects of gamma irradiation (see Huckins, 1978; Meistrich, 1982). Generally, these designs involve exposing the testis to cytotoxic doses of radiation, and waiting for 3–5 weeks for the radioresistant stem cells to begin to repopulate the tubules. These studies are clearly designed to circumvent morphologic evaluations, which are difficult to perform accurately on spermatogonia, because of their low numbers and infrequent divisions of the early stages. These same approaches may be useful in assessing the effects of highly toxic chemicals.

The foregoing has provided some examples of compounds that affect selected populations of cells in the testis. The frequency at which these effects are seen in a section of testis depends on the affected cell. For example, Sertoli cell toxicants like 2,5-hexanedione and diethylhexyl phthalate tend to affect tubules in many, but not all, stages of spermatogenesis; thus, the lesion is fairly widespread throughout the section. Conversely, the EGME lesion is quite specific, and at early times, only those tubules in stage XIV show signs of cell death, primarily in spermatocytes. Similarly, a lesion that appears to affect the fate of residual bodies will only appear in those stages where residual bodies are present (stages VIII and IX, and to a lesser extent, stage X).

In addition, different cells may manifest necrosis differently. At the LM level, degenerating spermatids undergo nuclear chromatolysis, leaving a peripheral ring of nucleoplasm—thus the term "ring-like nuclei" (Wong and Hruban, 1972). Late spermatids can lose cytoplasm; these cytoplasmic blebs show up in the lumina of the tubules as round PAS-positive "droplets." Spermatocytes that have been affected by treatment have a different staining pattern in the cytoplasm or nucleus (see Chowdhury and Steinberger, 1964), and Sertoli cells frequently have "vacuoles" often deriving from the smooth endoplasmic reticulum. These "vacuoles" can take several forms, and are produced by many different treatments (Kierszenbaum and Mancini, 1973; Hausler and Hodel, 1979; Maxwell, 1981). Their presence is generally taken as indication of a Sertoli cell effect, but their relation to the mechanism of action of these different compounds is unknown.

Given these types of changes, it is easy to determine the "target cell" for a compound or treatment, where a "target cell" is defined as the cell that shows the first observable signs of structural change. It is crucial, when considering target cells, to remember that the first visible change may not be the same as the site of the initial biochemical lesion. For example, a germ cell may degenerate due to a primary effect in a Sertoli cell. It is less easy to determine the target cell in a situation where the affected process is the interaction between two cell types. Such a process is the movement of spermatid cytoplasm from the late-step spermatid to the Sertoli cell via the tubulobulbar complex (Russell, 1979), a process that is visibly affected by treatment with methyl chloride (Chapin *et al.,* 1984d). It is difficult to determine in such cases which cell is principally affected, the spermatid or the Sertoli cell. Careful examination of sections in the EM may reveal the lack of such tubulobulbar complexes, but such studies are time-consuming and open to question.

The "blood–testis barrier" is really a blood–tubule barrier, formed by tight junctions between adjacent Sertoli cells. These tight junctions separate the epithelium into a basal compartment, which communicates freely with the interstitial space, and a luminal compartment, which is closed to the peritubular space. Lanthanum tracer molecules penetrate the epithelium to surround the spermatogonia and early spermatocytes that remain in the basal compartment, but the tracer cannot penetrate past these tight junctions (Dym and Fawcett, 1970). These junctions can be disrupted by ligating the efferent ducts, which causes the intraluminal pressure of the tubules to rise. It has been occasionally postulated that this "barrier" might be a target for toxicant action in the testis. However, the Sertoli cells of men with testicular disorders and aspermatogenesis have been found to have normal junctional complexes (de Kretser *et al.,* 1981); these complexes are similarly unchanged after cryptorchidism, hypomagnesia, experimental uremia, and dibromochloropropane (DBCP) exposure (see Aumuller *et al.,* 1980, 1981; Maxwell, 1981; and Biava *et al.,* 1978, respectively). However, a recent report has found temporary disruption of these tight junctions after cisplatin treatment (Pogatch *et al.,* 1986), suggesting that this should be more carefully evaluated for other compounds.

V. Functional Correlates of Pathology

Finally, there is the question of what these morphologic changes mean to the fertility of the animal: do pathologic changes in the testis result in reduced fertility? The answer is undoubtedly "yes," and the literature has a number of examples that support this contention (DBCP, Kluwe *et al.,* 1983a,b; dimethyl methylphosphonate, Dunnick *et al.,* 1984; for methyl chloride, Chapin *et al.,* 1984d, and Working *et al.,* 1985a,b; EGME, Chapin *et al.,* 1985a,b). Indeed, it is one of the fundamental assumptions of pathology that modified structure will, if sufficient in extent and/or duration, result in modified function (ability to reproduce). There are numerous examples of this in many mammalian species, including humans.

How much structural change is required to affect fertility depends on the nature of the change and the mechanism by which it is produced and on the time that fertility is measured after the onset of dosing. Compounds that affect only the mitochondrial function of epididymal spermatozoa will produce little change in the testis, but have a dramatic effect on sperm motility and therefore fertility. Conversely, it is quite possible to produce significant testicular histopathology without affecting fertility, at least until the epididymis is depleted of normal sperm and is full of defective forms resulting from treatment (Chapin *et al.,* 1985a,b). Overall, while there is a generally good correlation between histopathology and fertility, the precise degree of this correlation for an effect short of complete epithelial destruction cannot be predicted *a priori.*

VI. Summary

This chapter should have helped the reader become convinced that very good preservation of the structure of the testis is desirable and easily accomplished. It has briefly discussed some of the available techniques, and has shown how different techniques lead to different final products, some much better and more useful than others. The histochemical and immunohistochemical sections are meant to suggest that such methodologies, or similar ones, might be applied to particular problems. The discussion then moved to some examples of studies that identified specific cells as targets for treatment in the testis: some cell types show toxicity fairly readily, while others require different approaches to determine whether they are affected. The references to this section also illustrated different ways that dosing can be conducted, and documented how dosing for a very limited time and observing the sequelae allows the investigator to pinpoint the vulnerable cell(s). The discussion closed by noting that, generally, fertility correlates well with the structure of the testis, but that this depends both on the length of dosing and on the time that mating was performed relative to when dosing was initiated.

The overall message, then, is that efforts that the investigator makes to improve the quality of material for testicular histopathlogy will be richly rewarded, and when combined with a study design that allows for the kinetics of spermatogenesis, a great deal of information can be obtained about where in the testis a compound acts. Such methods are easily within reach of any investigator.

References

Aumuller, G., Hartmann, K., Giers, U., and Schenk, B. (1980). Fine structure of the Sertoli cells of the rat testis in experimental unilateral cryptorchidism. *Int. J. Androl.* **3,** 301–311.

Aumuller, G., Ritz, E., Kreusser, E., and Spiegelberg, U. (1981). Ultrastructure of the rat testis in experimental uremia. *Andrologia* **13,** 163–173.

Bergh, A. (1982). Local differences in Leydig cell morphology in the adult rat testis: Evidence for a local control of Leydig cells by adjacent seminiferous tubules. *Int. J. Androl.* **5,** 325–330.

Biava, C. G., Smuckler, E. A., and Whorton, D. (1978). The testicular morphology of individuals exposed to dibromochloropropane. *Exp. Mol. Pathol.* **29,** 448–458.

Bogdanffy, M. S., Randall, H. W., and Morgan, K. T. (1985). Cellular localization of aldehyde dehydrogenase activity by histochemistry in glycol methacrylate embedded tissues. *Toxicologist* **5,** 176.

Bogdanffy, M. S., Randall, H. W., and Morgan, K. T. (1986). Histochemical localization of aldehyde dehydrogenase in the respiratory tract of the Fischer 344 rat. *Toxicol. Appl. Pharmacol.* **82,** 560–567.

Breitnecker, G., Bock, P., and Lunglmayr, G. (1974). Histochemical localization of phosphohydrolases in testes of man and rat. *Z. Anat. Entwickl.-Gesch.* **143,** 301–313.

Brinn, N. T., and Pickett, J. P. (1979). Glycol methacrylate for routine, special stains, histochemistry, enzyme histochemistry, and immunohistochemistry. *J. Histotechnol.* **2,** 125–130.

Cameron, D. F., and Snydle, F. E. (1985). Selected enzyme histochemistry of Sertoli cells. 1. Immature rat Sertoli cells *in vitro. Andrologia* **17,** 64–71.

Cater, B. R., Cook, M. W., Gangolli, S. D., and Grasso, P. (1977). Studies on dibutyl phthalate-induced testicular atrophy in the rat: Effect on zinc metabolism. *Toxicol. Appl. Pharmacol.* **41,** 609–618.

Chapin, R. E., Morgan, K. T., and Bus, J. S. (1983). The morphogenesis of testicular degeneration induced in rats by orally administered 2,5-hexanedione. *Exp. Mol. Pathol.* **38,** 149–169.

Chapin, R. E., Dutton, S. L., Ross, M. D., Sumrell, B. M., and Lamb, J. C., IV. (1984a). Development of reproductive tract lesions in male F344 rats after treatment with dimethyl methylphosphonate. *Exp. Mol. Pathol.* **41,** 126–140.

Chapin, R. E., Ross, M. D., and Lamb, J. C., IV. (1984b). Immersion fixation methods for glycol methacrylate-embedded testes. *Toxicol. Pathol.* **12,** 221–227.

Chapin, R. E., Dutton, S. L., Ross, M. D., Sumrell, B. M., and Lamb, J. C., IV. (1984c). The effects of ethylene glycol monomethyl ether on testicular histology in F344 rats. *J. Androl.* **5,** 369–380.

Chapin, R. E., White, R. D., Morgan, K. T., and Bus, J. S. (1984d). Studies of lesions induced in the testis and epididymis of F344 rats by inhaled methyl chloride. *Toxicol. Appl Pharmacol.* **76,** 328–343.

Chapin, R. E., Dutton, S. L., Ross, M. D., and Lamb, J. C., IV. (1985a). Effects of ethylene glycol monomethyl ether on mating performance and epididymal sperm parameters in F344 rats. *Fundam. Appl. Toxicol.* **5,** 182–189.

Chapin, R. E., Dutton, S. L., Ross, M. D., Swaisgood, R. R., and Lamb, J. C., IV. (1985b). The recovery of the testis over 8 weeks after short-term dosing with ethylene glycol monomethyl ether: Histology, cell-specific enzymes, and rete testis fluid protein. *Fundam. Appl. Toxicol.* **5,** 515–525.

Chowdhury, A. K., and Steinberger, E. (1964). A quantitative study of the effect of heat on germinal epithelium of rat testes. *Am. J. Anat.* **115,** 509–524.

Clermont, Y., Leblond, C. P., and Messier, B. (1959). Duree du cycle de l'epithelium seminal du rat. *Arch. Anat. Microscop. Morphol. Exp. Suppl.* **48,** 37–56.

Creasy, D. M., and Foster, P. M. D. (1984). The morphological development of glycol ether-induced testicular atrophy in the rat. *Exp. Mol. Pathol.* **40,** 169–176.

Creasy, D. M., Foster, J. R., and Foster, P. M. D. (1983). The morphological development of di-*n*-pentyl phthalate-induced testicular atrophy in the rat. *J. Pathol.* **139,** 309–321.

Curto, K. A., and Thomas, J. A. (1982). Comparative effects of diethylhexyl phthalate or monoethylhexyl phthalate on male mouse and rat reproductive organs. *Toxicol. Appl. Pharmacol.* **62,** 121–125.

de Kretser, D. M., Kerr, J. B., and Paulsen, C. A. (1981). Evaluation of the ultrastructural changes in the human Sertoli cell in testicular disorders and the relationship of the changes to the levels of serum FSH. *Int. J. Androl.* **4,** 129–144.

Dunnick, J. K., Gupta, B. N., Harris, M. W., and Lamb, J. C., IV. (1984). Reproductive toxicity of dimethyl methylphosphonate in the male Fischer 344 rat. *Toxicol. Appl. Pharmacol.* **72,** 379–387.

Dym, M., and Fawcett, D. W. (1970). The blood–testis barrier in the rat and the physiological compartmentation of the seminiferous epithelium. *Biol. Reprod.* **3,** 308–326.

Ewing, L. L., Zirkin, B. R., and Chubb, C. (1981). Assessment of testicular testosterone production and Leydig cell structure. *Environ. Health Perspect.* **38,** 19–27.

Fawcett, D. W. (1975). Ultrastructure and function of the Sertoli cell. *In* "Handbook of Physiology," Vol 5 (D. W. Hamilton and R. O. Greep, eds.), pp. 21–56. American Physiological Society, Washington, D.C.

Fox, C. H., Johnson, F. B., Whiting, J., and Roller, P. P. (1985). Formaldehyde fixation. *J. Histochem. Cytochem.* **33,** 845–853.

Hagenas, L., Ploen, L., and Ritzen, E. M. (1978). The effect of nitrofurazone on the endocrine, secretory, and spermatogenic functions of the rat testis. *Andrologia* **10,** 107–126.

Hards, D. K. (1983). Stain removal from methacrylate sections. *Med. Lab. Sci.* **40,** 393–394.

Hashimoto, K., Sakamoto, J., and Tanii, H. (1981). Neurotoxicity of acrylamide and related compounds and their effects on male gonads in mice. *Arch. Toxicol.* **47,** 179–189.

Hausler, A., and Hodel, C. (1979). Ultrastructural alterations induced by two different antispermatigenic agents in the seminiferous epithelium of rat testes. *Arch. Toxicol. Suppl.* **2,** 387–392.

Huckins, C. (1978). *Biol. Reprod.* **19,** 747–760.

Kierszenbaum, A. L., and Mancini, R. E. (1973). Structural changes manifested by Sertoli cells during experimental allergic orchitis in guinea pigs. *J. Reprod. Fertil.* **33,** 119–122.

Kluwe, W. M., Lamb, J. C., IV, Greenwell, A. E., and Harrington, F. W. (1983a). 1,2-Dibromo-3-chloropropane-induced infertility in male rats mediated by a post-testicular effect. *Toxicol. Appl. Pharmacol.* **71,** 294–298.

Kluwe, W. M., Gupta, B. N., and Lamb, J. C., IV. (1983b). The comparative effects of 1,2-dibromo-3-chloropropane and its metabolites, 3-chloro-1,2-propaneoxide (epichlorhydrin), 3-chloro-1,2-propanediol (alphachlorhydrin), and oxalic acid, on the urogenital system of male rats. *Toxicol. Appl. Pharmacol.* **70,** 67–86.

Kormano, M., and Hovatta, O. (1972). Contractility and histochemistry of the myoid cell layer of the rat seminiferous tubules during postnatal development. *Z. Anat. Entwickl.-Gesch.* **137,** 239–248.

Kuehn-Velten, N., Wolff, J., Passia, D., and Staib, W. (1982). Determination of non-specific esterase activity in rat testis interstitial cells: A marker for evaluation of a Leydig cell purification procedure and of cell viability during incubation and continuous superfusion. *Cell Mol. Biol.* **28,** 473–482.

Lamb, J. C., IV, and Chapin, R. E. (1985). Experimental models of male reproductive toxicology. *In* "Endocrine Toxicology" (J. A. Thomas, K. S. Korach, and J. A. McLachlan, eds.), pp. 85–115. Raven, New York.

Leblond, C. P., and Clermont, Y. (1952). Definition of the stages of the cycle of the seminiferous epithelium of the rat. *Ann. N.Y. Acad. Sci.* **55,** 548–573.

Lee, I. P., and Gillies, P. J. (1984). Ultrastructural alterations in hexafluoroacetone-induced testicular atrophy in the rat. *Exp. Mol. Pathol.* **40,** 29–37.

Maxwell, W. L. (1981). Ultrastructure of the seminiferous tubules in magnesium-deficient rats. *Arch. Androl.* **6,** 13–25.

Mazzocchi, G., Robba, C., Rebuffat, P., Gottardo, G., and Nussdorfer, G. G. (1982). Effects of a chronic treatment with testosterone on the morphology of the interstitial cells of the rat testis: an ultrastructural, stereological study. *Int. J. Androl.* **5,** 130–136.

Meistrich, M. L. (1982). Quantitative correlation between testicular stem cell survival, sperm production, and fertility in the mouse after treatment with different cytotoxic agents. *J. Androl.* **3,** 58–68.

Namba, M., Dannenberg, A. M., and Tanaka, F. (1983). Improvement in the histochemical demonstration of acid phosphatase, beta-galactosidase, and non-specific esterase in glycol methacrylate tissue sections by cold temperature embedding. *Stain Technol.* **58,** 207–213.

Pedraza, M. A., Mason, D., Doslu, F. A., Marsh, R. A., and Boblett, J. P. (1984). Immunoperoxidase methods with plastic-embedded materials. *Lab. Med.* **15,** 113–115.

Pogatch, L. M., Huang, H. F. S., Lee, Y., and Giglio, W. (1986). Sertoli cell function in cisplatin-treated Sprague Dawley rats. *J. Androl.* **7,** 28P.

Puchtler, H., and Meloan, S. N. (1985). On the chemistry of formaldehyde fixation and its effects on immunohistochemical reactions. *Histochemistry* **82,** 201–204.

Redi, C. A., Hilscher, B., and Winkling, H. (1983). Stage-dependent enzymatic activities in spermatogenesis of mice with the standard karyotype and of chromosomal variants with impaired fertility. *Andrologia* **15,** 322–330.

Roosen-Runge, E. C. (1973). Germinal cell loss in normal metazoan spermatogenesis. *J. Reprod. Fertil.* **35,** 339–348.

Russell, L. D. (1979). Spermatid–Sertoli tubulobulbar complexes as devices for elimination of cytoplasm from the head region of late spermatids of the rat. *Anat. Rec.* **194,** 233–246.

Russell, L. D., and Clermont, Y. (1977). Degeneration of germ cells in normal, hypophysectomized, and hormone-treated rats. *Anat. Rec.* **187,** 347–366.

Russell, L. D. (1983). Normal testicular structure and methods of evaluation under experimental and disruptive conditions. *In* "Reproductive and Developmental Toxicity of Metals" (T. W. Clarkson, G. F. Nordberg, and P. R. Sawyer, eds.), pp. 227–252. Plenum, New York.

Russell, L. D., Malone, J. P., and Karpas, S. L. (1981). Morphological patterns elicited by agents affecting spermatogenesis by disruption of its hormonal stimulation. *Tissue Cell* **13,** 369–380.

Russell, L. D., Lee, I. P., Ettlin, R., and Malone, J. P. (1983). Morphological pattern of response after administration of procarbazine: Alteration of specific cell associations during the cycle of the seminiferous epithelium of the rat. *Tissue Cell* **15,** 391–404.

Scott, J. E., and Persaud, T. V. N. (1977). Morphological and enzyme histochemical changes in the reproductive tract of the male rat induced by acetylsalicylic acid. *Acta Histochem.* **60,** S228–246.

Sever, L. E., and Hessol, N. A. (1985). Toxic effects of occupational and environmental chemicals on the testes. *In* "Endocrine Toxicology" (J. A. Thomas, K. S. Korach, and J. A. McLachlan, eds.), pp. 211–248. Raven, New York.

Somkuti, S. G., Lapadula, D. M., Chapin, R. E., Lamb, J. C., IV, and Abou-Donia, M. B. (1986). Testicular toxicity of tri-o-cresyl phosphate in F344 rats. *Toxicologist* **6,** 290.

Steinberger, E., and Steinberger, A. (1975). Spermatogenic function of the testis. *In* "Handbook of Physiology," Vol. 5 (D. W. Hamilton and R. O. Greep, eds.), pp. 1–22. American Physiological Society, Washington, D.C.

Steinberger, E., and Tjioe, D. Y. (1969). Spermatogenesis in rat testes after experimental ischemia. *Fertil. Steril.* **20,** 639–649.

Waller, D. P., Killinger, J. M., and Zaneveld, L. J. D. (1985). Physiology and toxicology of the male reproductive tract. *In* "Endocrine Toxicology" (J. A. Thomas, K. S. Korach, and J. A. McLachlan, eds.), pp. 269–333. Raven, New York.

Wong, T. W., and Hruban, Z. (1972). Testicular degeneration and necrosis induced by chlorcyclazine. *Lab. Invest.* **26,** 278–289.

Working, P. K., Bus, J. S., and Hamm, T. E., Jr. (1985a). Reproductive effects of inhaled methyl chloride. 1. Mating performance and dominant lethal assay. *Toxicol. Appl. Pharmacol.* **77,** 133–143.

Working, P. K., Bus, J. S., and Hamm, T. E., Jr. (1985b). Reproductive effects of inhaled methyl chloride. 2. Spermatogonial toxicity and sperm quality. *Toxicol. Appl. Pharmacol.* **77,** 144–157.

Chapter 8

Evaluation of Copulatory Behavior and Sperm in Rats: Role in Reproductive Risk Assessment*†

HAROLD ZENICK

U.S. Environmental Protection Agency
Reproductive Effects Assessment Group (RD689)
Office of Health and Environmental Assessment
Washington, D.C. 20460

HELEN GOEDEN

Department of Pharmacology and Therapeutics
University of Calgary
Calgary, Alberta, Canada

*The information described in this chapter has not been subjected to the Agency's peer and administrative review and therefore does not necessarily reflect the views of the Agency and no official endorsement should be inferred.

†This work was supported by Reproductive Hazards in the Workplace Research Grants 15-58 and 15-59 from the March of Dimes Research Foundation, NIOSH RO1-OHO1272, EPA CR808880, and NIEHS ES07073.

I. Introduction

The purpose of this chapter is to examine the role that the assessments of copulatory behavior and sperm have in reproductive risk assessment. This text compliments the coverage of breeding and fertility testing and morphological evaluations of the testes provided in Chapters 6 and 7 in this volume. The commentary is almost exclusively focused on the use of the rat as an animal model.

Efforts in the area of risk assessment have concentrated primarily on cancer as an outcome. However, attention is now being directed toward the development of stategies for assessing risk to other target systems. The male reproductive system is a prime candidate for such efforts. Research conducted over the past decade has shown that the germ cell may be extremely sensitive to chemical injury, often in the absence of other apparent clinical signs. One needs only to look at the data on dibromochloropropane and several of the monoalkyl glycol ethers to have ample support for this statement. Moreover, the quality and quantity of sperm production by the human testes appear marginal relative to other species. Thus, the reproductive capability of humans may be more readily compromised by chemical exposure than other mammalian species.

As toxicologists, much of our effort must be directed toward developing and evaluating methodologies and endpoints that will assist in the identification of male reproductive toxicants and subsequent dose–response quantification. The assessments of copulatory behavior and sperm have obvious roles to play in this process.

II. Traditional Endpoints in Male Reproductive Risk Assessment

Identification of male reproductive hazards has traditionally relied on results of acute and subchronic toxicity testing and breeding studies. The endpoints usually obtained from such investigations may include organ weights, histopathology, mating success, and measures of fertility and fetal outcome.

A. Organ Weights and Histopathological Evaluations

Organ weights are usually obtained irrespective of the study protocol. For the male, these routinely include the testes and, less commonly, the epididymis, vas deferens, and accessory sex glands. Testes weight can be measured easily and accurately and demonstrates modest variation among members of a given test species (Table I). This low variability suggests that testicular

TABLE I

Relative Sensitivity of Testicular and Epididymal Sperm Parameters[a]

Parameter	Coefficient of variation[b]	Percent of difference detected		
		$N = 10$	$N = 15$	$N = 20$
Testis weight	6.7 (4.65)	8.77	7.16	6.20
Epididymal weight	10.2 (9.40)	13.35	10.89	9.43
Sperm rate production[c]	14.9 (16.56)	19.50	15.91	13.77
Sperm count per gram epididymis	23.4 (29.32)	30.36	24.99	21.65
Percent normal morphology	2.7	3.53	2.88	2.49
Percent motile	24.4 (16.0)	31.49	25.64	22.22

[a]Represents data on control values derived from four to five studies conducted in this laboratory on Long–Evans rats.

[b]Numbers in parentheses represent values reported by Blazak *et al.* (1985) for Fischer 344 rats.

[c]Sperm rate production = spermatid enumeration/rat.

weight would be a sensitive marker of gonadal injury; however, it is often insensitive. Substantial damage to the testes may be required before a significant change in weight can be detected. This contradiction may arise from several factors, including the contribution of other nongerminal elements to testicular weight (e.g., interstitial cells) and the early physiological/biochemical reactions to injury that may actually increase testicular weight (e.g., edema, inflammation, and cellular infiltration). The absence of a meaningful correlation between daily sperm production and testes weight also argues against its sensitivity as an early predictor of gonadal toxicity (Robb *et al.*, 1978).

Histopathological data on the testes are also routinely obtained. This aspect of reproductive assessment is reviewed in Chapter 7; thus, only a few observations are offered here.

As most commonly practiced, the routine procedures of tissue processing (i.e., formalin fixation and paraffin embedding) produce a number of artifacts and allow for the detection of only gross damage. With the utilization of more appropriate preparative techniques and more precise quantitative and qualitative methods, histological analysis can provide a powerful tool for the detection of the initial site of lesion and time of onset of damage. Such information is invaluable for specifying the cell type to be evaluated in subsequent mechanistic studies. The existence of similar mechanisms in humans strengthens any interspecies comparison that may be drawn. In terms of risk assessment, there are two major limitations of these methods. First is the inability to predict the functional significance of modest lesions; second is the improbability of applying such procedures to a human population to obtain direct verification of laboratory findings and vice versa.

B. Breeding Studies

Breeding studies have served as the other major source of data on male reproductive toxicants. The evaluation of fertility and the resulting pregnancy outcomes provide insight as to the functional consequence of reproductive injury. Fertility tests are unique in being a simultaneous test of the integrity and the interaction of a large number of component systems [e.g., central nervous system (CNS), endocrine, testes, etc.] The major breeding protocols include the Food and Drug Administration (FDA) Segment 1 test for drugs, the single and multigenerational tests used by the Environmental Protection Agency (EPA) and FDA (pesticides and food additives), and the dominant lethal test to assess genetic damage that may produce adverse pregnancy outcomes. Several limitations exist with the current protocols.

All of these protocols utilize the presence of copulatory plugs and/or a sperm-positive vaginal lavage to assume that copulatory behavior is unimpaired. The fallacy of such an assumption is discussed in Section III. A second limitation is that some of the protocols entail cohabitation of exposed male and exposed female animals, so that paternal contributions may be difficult to delineate without further studies. Moreover, the rate of successful mating for control males may vary widely across studies (Schwetz *et al.*, 1980), necessitating very large sample sizes. Breeding rates can be improved if one not only employs experienced male breeders, but also specifically mates them with proestrous females selected from the animal colony.

As is true of histopathological findings, animal data on fertility success (or the lack thereof) are difficult to compare to human populations, which are characterized by low rates of conception. Moreover, the couple must be treated demographically as the unit of analysis, making it more difficult to ascribe reproductive failure solely to the male.

An issue of greater concern relates to the insensitivity of fertility measures in laboratory animals to reflect reproductive injury. Normal males of most test species produce sperm in numbers that well exceed the minimal requirements for fertility (Amann, 1981). A substantial degree of gonadal damage may be tolerated without compromising fertility success. In some rodent strains, sperm production may be reduced by 90% without compromising fertility [e.g., C3H mice (Meistrich, 1982); Sprague-Dawley rats (Robaire, *et al.*, 1984)]. However, far less severe effects can have dramatic consequences in the man, who functions nearer to threshold in terms of the number of sperm needed to ensure reproductive competence. The outcome of this species difference is that negative data from a fertility study cannot be the sole basis for concluding that the test agent poses no reproductive hazard to men. Data on other end points may be necessary to verify or refute this position. Inclusion of data on copulatory behavior and sperm can provide more data on which to base decisions.

III. Expanding the Data Base—Evaluation of Copulatory Behavior

Sexual behavior is a very complicated phenomenon involving neural and endocrinological involvement of the central and peripheral nervous systems. For humans, a complex constellation of personality, social, and experiential factors also influence the initiation and performance of these behaviors. The existence of similar factors in other species is uncertain. Thus, a degree of uncertainty exists as to the extrapolation of animal data to humans for these endpoints. However, the perturbation of sexual behavior in animals can suggest the potential for effects in humans. Consistent with this position is data on known CNS agents that have been shown to disrupt sexual behavior in both animals and humans (Waller *et al.*, 1985; Rubin and Henson, 1979).

Several excellent reviews are available on the mechanisms controlling the development and performance of sexual behavior in rats. Rather than reiterate that literature, we have chosen to describe the methodological considerations in monitoring copulatory behavior and to provide examples of environmental agents demonstrated to perturb these behaviors.

A. Standard Laboratory Approach and Its Limitations

Although sexual behavior can be readily quantified in most test species, no direct evaluations of this behavior are performed in most reproductive toxicology studies. Rather, the presence of copulatory plugs or sperm-positive vaginal lavages has been taken as indirect evidence of successful mating. However, the presence of a seminal plug or sperm in the vagina does not confirm that the male performance was adequate to ensure sperm deposition and subsequent transport in the uterus. Disruption of the sexual dynamics of the male may not only impede ejaculation, but also result in inadequate sexual stimulation of the female. The degree of sexual preparedness of the female partner is critical to subsequent sperm transport in her genital tract.

In such circumstances, the "positive" female would fail to deliver a little. Under current protocols, such a "mating" failure would be reflected in the calculated fertility index and could erroneously be classified as a spermatotoxic effect.

There are other aspects of current breeding protocols that may serve to actually mask a decline in fertility potential of the male. As currently structured, the male and female cohabitate for several days (possibly for as long as 3 weeks). As a result, the opportunity exists for multiple sexual bouts during estrus, and several estruses during this period, which could compensate for impaired sexual behavior. Thus, an effect that could be of consequence to humans would go undetected in such a mating trial.

As additional factor that must be considered is the developmental stage at which exposure occurred to the male. Prenatal and/or prepubertal exposures may alter the central neural and peripheral somatic structures pivotal to the expression of masculinity and male behaviors. Adolescent or adult exposure may alter sexual arousal and initiative (Goy *et al.*, 1983). It is likely that most of the changes that accompany adult exposure are reversible; however, the reversibility of changes associated with early developmental exposure is less likely.

B. Systematic Evaluation of Copulatory Behavior

Evaluation of copulatory behavior should occur during the dark portion of the rodent's light–dark cycle. Maintenance of the colony on a reverse light–dark cycle will readily facilitate such efforts. Observations should take place in a Plexiglas cage to which the male has previously been adapted. Measures of copulatory function are expressed as the frequency of a given event and elapsed time between events (latency). Frequently employed measures include the following: mount latency, the latency between introduction of the female and the initiation of the first mount by the male; ejaculation latency, the latency between first mount and ejaculation; number of mounts, mounting the female without penile intromission; and number of intromissions, mounting with penile intromission. This latter act is accompanied by a vigorous pelvic thrust followed by back-pedalling and genital grooming. The act of ejaculation is apparent in that, when the male ejaculates, there is not only a vigorous thrust of the hindlimbs, but upon disengagement the male becomes refractory to the female's advances for several minutes. This interval well exceeds that occurring between mounts or intromissions. The presence of vaginal sperm can also confirm that ejaculation has occurred. With experience, these behaviors are easily distinguished and exhibit high interrater reliability ($r \geq 0.90$).

Other measures of copulation can be recorded, including the interval between mounts, the interval between intromissions, and the interval between ejaculation and the initiation of the next bout. However, the recording of these additional variables greatly increases the demand on the observer. A keyboard, interfaced to a computer, can facilitate both data acquisition and subsequent analyses.

If copulatory behavior is to be evaluated, sexually experienced males must be used. Sexually naive males are notoriously poor at copulating, requiring a prolonged time to ejaculate and then not necessarily ejaculating into the vagina. The performance is even poorer if the males are older at the time of the first mating trial (e.g., 6–8 months of age). Ironically, the design for many of the breeding protocols involves the use of such naive males, which may explain, in part, the high background (control) rate of mating failure (Schwetz *et al.*, 1980). Similarly, in dominant lethal studies, the high background levels of "apparent" preimplantation loss that are reported early in the course of serial mating may be a reflection of mating incompetence.

Our strategy has been to receive males into the laboratory at 70–80 days and, during their 2- to 3-week acclimatization period, provide them with several mating experiences. This can be done very efficiently by using ovariectomized, hormonally primed females (Zenick *et al.*, 1984a). Such females can be used repeatedly within and across sessions, and need only to be placed into the male's home cage. Each male is then evaluated for copulatory competence (in Plexiglas cages) prior to the initiation of exposure. Males failing to initiate a mount within 10 min or to ejaculate within 30 min are eliminated from the study. These data may also be used to establish a preexposure baseline against which subsequent evaluations of the male during treatment can be contrasted. The 10- and 30-min criteria are again applied to terminate a given session during the treatment phase of the study.

C. *Recent Studies on Environmental Agents That Disrupt Male Copulatory Behavior*

Studies have been conducted recently on the reproductive evaluations of three known neurotoxicants, namely, carbon disulfide (Tepe and Zenick, 1984; Zenick *et al.*, 1984b), trichloroethylene (Zenick *et al.*, 1984c), and acrylamide (Zenick *et al.*, 1986). In the course of these studies, all three agents were shown to impair copulatory behavior, and, in one case (Zenick *et al.*, 1986), produce a direct, adverse impact on fertility.

Exposure to carbon disulfide for 10 weeks (600 ppm, inhalation) produced a decrease in mount and ejaculation latencies as well as a decrease in ejaculated sperm count. However, the decline in the latter variable was not due to gonadal toxicity, since epididymal sperm counts were unaffected (Zenick *et al.*, 1984b). A more likely hypothesis is that the abbreviated copulatory bout (analogous to premature ejaculation?) may have affected the extent of sexual arousal in both partners and, in turn, the number of sperm deposited and transported in the female tract. Chester and Zucker (1970) have shown that disruptions in copulatory behavior can influence the number of sperm recovered from the female tract. Although fertility was not assessed in this study, the potential for subfertility exists in light of the reduced number of sperm in the uterus.

Studies of trichloroethylene (TCE) have shown that it produces an increase in ejaculation latency in the absence of other signs of gonadal toxicity (Zenick *et al.*, 1984c). The mechanism is not associated with TCE's anesthetic effects, for the males remain active throughout the copulatory bout, repeatedly initiating contact with the female. Rather, the increased latency appears to be the result of an opioid involvement. In support of this hypothesis, we have shown that the acute TCE prolongation of copulatory behavior is blocked by naloxone, an opiate antagonist, but not by quaternary naloxone; tolerance develops with prolonged exposure; and there was a cross tolerance with morphine (Nelson and Zenick, 1986). The effects of these copulatory impairments on fertility potential remain to be assessed. Moreover, the interaction of TCE with other environmental exposures such as alcohol or narcotics is also in need of evaluation. The possibility exists that such interactions would be synergistic and could lower the dose–response threshold for this TCE-induced effect. The probability of such combined exposures in the human population is significant.

Acrylamide (ACR) also produces disruptions in copulatory behavior in rats (Zenick *et al.*, 1986). Although the treated males can readily mount the females, they have difficulty in completing the more coordinated act of intromission. There are two consequences of such perturbations. First, sperm deposition may be hindered; second, the female may not receive adequate stimulation to facilitate subsequent sperm transport. Supporting evidence appears in both semen evaluations and fertility tests conducted on males receiving the highest dose of ACR (100 ppm in drinking water for 11 weeks). The majority of females mated to these treated males fail to have semen present in their uteri, despite the presence of copulatory plugs. In the fertility segment, plus incidence is again 100%, yet the pregnancy rate is only 33%. These data reinforce the view that disturbances in copulatory behavior can directly impact on rodent fertility.

Direct evaluation of sexual behavior is not warranted for all suspected reproductive toxicants. Likely candidates may be agents reported to exert neurotoxic effects (sensory and/or motor, peripheral and/or central). Several neurotoxicants have been reported to produce gonadal toxicity, including hexane, Kepone, acrylamide, and methyl chloride. Chemicals possessing androgenic or estrogenic properties also merit consideration [e.g., polychlorylbiphenyls (PCBs), diethylstilbestrol (DES), chlorinated hydrocarbon pesticides].

IV. Expanding the Data Base—Sperm Evaluations in the Rat

A major strength in conducting sperm evaluations in animals is that similar data can be obtained in humans, enhancing the ability to confirm laboratory observations and vice versa. The end points do not suffer from some of the methodological and statistical constraints associated with monitoring reproductive outcomes in human populations (e.g., reduced fertility, spontaneous abortion). In addition, a more direct interpretation can be offered as to the functional consequences of such alterations as contrasted with histological data. Wyrobek *et al.*, (1983a,b) have provided extensive summaries of the sperm effects of agents reported to be male reproductive toxicants.

Considerations were offered earlier on man's precarious status relative to other species in terms of sperm production rates. These present the best arguments for the inclusion of sperm evaluations in test protocols, rather than relying solely on the results from fertility assessments. Accurate estimates of spermatotoxicity can only be obtained through the conduct of sperm evaluations. The following sections review the approaches available for obtaining sperm samples and discuss the various spermatogenic endpoints that can be evaluated.

A. General Considerations

Sperm evaluations may be based either on samples from ejaculates, vas deferens, or cauda epididymis, or by counting homogenization-resistant spermatid nuclei recovered from testicular homogenates. The latter is an estimate of germ-cell production, maturation, and survival (Johnson *et al.*, 1980). In addition to sperm count, assessments of sperm motility and sperm morphology can be made on samples from ejaculates, vas deferens, or epididymis.

Depending on the level of quantification desired (see below), sperm assessments are relatively straightforward to conduct, not overly time-consuming, and can be implemented by a technician without extensive training. Evaluations of epididymal samples and testicular homogenates can be readily incorporated into pathology protocols employed in most toxicology studies.

Historically, the only laboratory species in which it was convenient to obtain ejaculates was the rabbit. However, recent strategies have been applied to evaluating serial ejaculates in the rat (Ratnasooriya *et al.*, 1980; Zenick *et al.*, 1984a). These approaches entail recovery of the semen sample from the genital tract of the female at a specified time interval postcopulation. Assessment of samples collected through electroejaculation is not recommended.

Evaluation of ejaculated samples provides advantages not possible with the other two approaches and provides the only semen source readily obtained from the human male. Data may be obtained from repeated evaluations on the same male over time, and may be more useful because of the marked interindividual variability that characterizes sperm count data. A longitudinal study design may have greater detection sensitivity and require smaller sample sizes (Wyrobek *et al.*, 1984). In addition, a pretreatment baseline may also be obtained on each male, against which changes during the course of treatment can be better delineated. The extent of recovery (reversibility) can also be accurately determined for a given individual.

B. Methodological Considerations in Handling Rat Sperm Samples

Rat spermatozoa are not routinely manipulated *in vitro*. One reason for this may be that rat spermatozoa are highly sensitive to a variety of environmental factors, such as temperature, pH, and dilution (Niwa and Chang, 1974). Quantitative analyses of sperm therefore require expediency and rigid control of environmental factors. We have found that utilization of a CO_2 incubator is not necessary as long as the assessment period is not greater than 10–15 min.

Throughout the analysis of functional sperm measures (e.g., motility and viability), the temperature of the sample should be maintained at 37°C. Spermatozoa will quickly lose motility as the temperature drops. The use of prewarmed syringes for sample collection, warm culture tubes maintained in a 37°C water bath, prewarmed microscope slides and coverslips, and an air-curtain incubator for the microscope stage will facilitate proper temperature control.

The optimal pH range for rat spermatozoa is relatively narrow (7.4–7.8). Basic salt solutions are unable to maintain the required pH range for any length of time. The addition of HEPES buffer to Toyoda and Chang's *in vitro* fertilization medium will provide a stable pH within the desired range (Paz-Frenkel *et al.*, 1978).

Sperm quality is also affected by sperm concentration (i.e., the degree of dilution of the sample). High concentrations will quickly deplete the medium of substrates as well as interfere with quantitative assessment of indices such as motility. Low concentrations (less than 1 million/ml) may produce poor quality for reasons that we have yet to identify. In our experience, sperm concentrations of 8–10 million/ml seem to be optimal for our analytical procedures.

C. Spermatogenic Endpoints

1. Sperm Count and Concentration

Sperm counts may be derived from ejaculated, vas deferens, epididymal, or testicular samples. For ejaculates, both sperm concentration (number of sperm per milliliter of ejaculate) and total sperm per ejaculate (sperm concentration × volume) should be analyzed. If the male is maintained on a schedule that provides for ejaculation or copulation at a standardized interval (see Amman, 1981), ejaculated sperm count (daily sperm output) can be used as an index of daily sperm production in the testes.

Epididymal evaluations should utilize sperm from only the caudal portion of the epididymis. The processing and evaluation of epididymal samples is rapid and straightforward. The landmark utilized in sectioning the cauda from the body of the epididymis will vary across laboratories. Thus, it has been customary to express sperm count on the basis of weight of the cauda epididymis. However, since sperm contribute directly to epididymal weight, expression of the data as a ratio may actually mask true declines in sperm number. Thus, data on absolute counts should also be analyzed.

Sperm production may also be estimated from the enumeration of spermatid nuclei recovered from testicular homogenates. These spermatid nuclei represent the elongated spermatids that are resistant to homogenization. The cell counts are a direct measure of both stem-cell division and germ-cell survival through the various proliferative and differentiating stages of spermatogenesis (Meistrich, 1982). As with epididymal samples, the methods for sample procurement and evaluation are simple and straightforward (Robb *et al.*, 1978). Spermatid enumeration provides an attractive alternative to the more laborious morphometric procedures for estimating sperm production, such as counting the number of stem cells in histological sections (Johnson *et al.*, 1980). Moreover, the latter gives no insight into survival of those cells during the course of spermatogenesis.

Spermatid enumerations have not been routinely employed as an index of reproductive toxicity. However, results of a recent study of lead-induced reproductive dysfunction (Sokol *et al.*, 1985) support the inclusion of such a measure into test protocols. Those authors were able to demonstrate a highly significant negative correlation ($r = 0.81$) between blood lead levels and spermatid counts in young adult rats treated with either 0.1% or 0.3% lead acetate for a 30-day period. The importance of this finding is that these effects were present at blood lead levels that are reported in the human population ($\sim 30-60\ \mu g/dl$). These findings are also consistent with reports of reproductive dysfunction in men occupationally exposed to lead (Lancranjan *et al.*, 1975; Cullen *et al.*, 1984).

Since the variability associated with spermatid counts is less than that seen with epididymal sperm counts (Table I), the ability to detect a decrease in sperm production should be enhanced by the use of spermatid counts. However, spermatid enumerations reflect the integrity of spermatogenic processes only in the testes. Posttesticular damage can only be detected through evaluations of ejaculated, vas deferens, or epididymal samples. Evaluations of motility, viability, fragility, and other functional properties of the sperm must also rely on analyses of epididymal or ejaculated samples.

Measures of sperm count have been the most frequently reported semen variable in both the animal and human literature (Wyrobek, 1983a,b). However, this measure is characterized by marked intra- and interindividual variability. Ejaculated sperm count is influenced by many extraneous variables, including the length of the abstinence period and the ability to collect the whole sample. Prospective, longitudinal sampling would facilitate the interpretation of a given change in sperm count for an individual. However, this opportunity is rarely afforded. Rather, interpretation must often be based on a single sample. When possible, additional spermatogenic measures should be assessed in conjunction with sperm count (motility, morphology) to better comprehend the significance of the obtained values.

2. Sperm Morphology

Sperm morphology refers to changes in head and/or tail shape that can be evaluated in either ejaculated, vas deferens, or epididymal samples. In the majority of animal studies, only head shape is evaluated: more attention should be paid to alterations in midpiece or tail morphology.

The traditional approach to characterizing morphology has relied on a subjective categorization of sperm shape from the examination of stained slides at the light microscopic level. Such an approach may be adequate for mice and rats, with their distinctly angular sperm head shapes. However, the heterogeneity of form and structure in human sperm makes it difficult for the morphologist to clearly define limits of normality. The subjective assessment of human sperm can be markedly improved with the utilization of a dimensionalized overlay (Katz *et al.*, 1982) or the employment of a systematic decision-tree process for categorization (Wyrobek *et al.*, 1984). The process can be further quantified by employing a variety of computer-linked digitizing systems to provide data on length, width, perimeter, and area of the sperm head.

Sperm morphology is quite stable and characteristic of an individual (and in animals, for a strain) over time. Of the various sperm measures, it has the least variability, enhancing its utility in the detection of spermatotoxic events (see Table I). However, alterations in other reproductive endpoints, such as motility or decreased fertility, may occur in the absence of an increase in abnormal sperm forms.

Two types of morphological alterations need to be distinguished. One is reflective of mutagenic effects, traditionally labeled teratospermia, while the other is an expression of cellular degeneration. Mutagen-induced alterations in sperm morphology have received the greatest attention (Wyrobek *et al.*, 1983a). However, the interrelationship between either type of morphological change and other sperm measures (e.g., motility, viability, and sperm count) has not been thoroughly examined. The relative dose–response relationship between alterations in these indices is also unknown.

The reproductive implications of abnormal sperm morphology are uncertain. The majority of studies in animals and humans have suggested that there is a marked decrease in the percentage of structurally abnormal cells within a population that reach the oviduct (e.g., Redi *et al.*, 1984; Nestor and Handel, 1984). This bias may be the result of an active selection process in the female tract, and/or of the cells being incapable of reaching the site of fertilization [e.g., differences in flagellar beat frequency (Katz *et al.*, 1982) or midpiece or tail defects (Jeulin *et al.*, 1985)]. Subfertility is the most likely outcome. Results from *in vitro* studies indicate that even if these abnormal cells gain access to the ova, fertilization and/or subsequent cell division are unlikely.

Exposure to toxic agents can result in heritable damage leading to abnormal sperm morphology in the progeny. Conversely, transmissible germ-cell mutations can occur in the absence of any warning indicator such as abnormal sperm morphology. Finally, the relationship between these alterations and other karyotypic changes remains uncertain (de Boer *et al.*, 1976).

3. Sperm Motility

The biochemical milieu in the testes and epididymis is highly regulated to insure the proper maturation and development of sperm and the acquisition of critical functions including motility. With chemical exposures, the perturbation of this homeostasis is likely. When such disruptions occur, alterations in sperm properties such as motility may occur at lower doses and/or earlier in the course of exposure than other spermatotoxic events (e.g., decreases in sperm count). Despite the potential sensitivity of such markers, few studies have examined motility as an endpoint (Wyrobek *et al.*, 1983a,b).

Motility estimates may be obtained from either ejaculated, vas deferens, or epididymal samples. Motility, like sperm count, is influenced by a number of variables, including abstinence and the elapsed time between obtaining the sample and evaluation. Moreover, the medium into which the sample is placed must be of a proper temperature, pH, and composition for that species. Evaluations of motility in animal or human studies require tight experimental control of these factors. Faulty laboratory procedures will severely compromise the integrity of the sperm, making it impossible to delineate treatment-related effects.

Motility has historically been evaluated utilizing subjective microscopic evaluations. Estimates of percent motile can be made and some rating of the quality of motility can be offered (i.e., the degree to which sperm show progressive, linear movement). However, these techniques are unreliable, showing high intra- and interrater variability. Moreover, no opportunity is provided to obtain a permanent record for subsequent referral or validation.

New techniques utilizing photomicrographic or videomicrographic approaches have been introduced (e.g., Katz and Overstreet, 1981). These strategies are far more objective, provide a permanent record, and allow for additional data to be obtained on the sperm. Several specific parameters have been developed to characterize rodent motility (Tessler and Olds-Clarke, 1985). For each sample, five indices can be measured: (1) curvilinear velocity, an index of swimming speed; (2) coefficient of variation of move length, an index of speed constancy; (3) progressive ratio, an index of all deviations in movement from a straight line; (4) linear index, an index of straightness (progressiveness); and (5) curvilinear progressiveness ratio, an index of the degree of lateral oscillation about the trajectory. These indices may be affected before an overall decrease in the percent of motile sperm is seen. With these techniques, data on individual sperm can be derived. Statistical analyses can then examine not only shifts in central tendencies but also distribution characteristics of that population of sperm. The increasing availability of computer-linked image-analyzing systems will markedly decrease the time required to perform such evaluations while greatly improving accuracy. However, no matter how sophisticated the analytical system, the burden still remains with the investigator to ensure the proper handling of the sample.

Alternative methods of estimating motility calculate the degree of displacement of various energy sources as the sperm swim through a given medium (e.g., light or photon scattering). Such approaches are highly accurate in deriving the relative percent motility for a population of cells; however, data on individual cells are lost. These methods also do not allow for reassessment of that population at a later date, since these activities can not be permanently recorded on film or videotape.

Evaluations of motility in initial studies may be limited to only characterizing the percent of motile cells in the population. The more informative indices such as swimming speeds or patterns could be evaluated in subsequent investigations. If motility has been recorded on film or videotape, then data on these measures could be gathered at any time.

4. Measures of Viability and the Acrosome Reaction

The fragility of rat sperm *in vitro* and the size and shape of the acrosome have, in the past, thwarted investigators' attempts to quantify viability and acrosomal integrity. The acrosome of the rat spermatozoa is very thin, and the acrosome reaction is rarely detectable under the light microscope. To assess acrosomal integrity and viability, we have modified the fluorescent acridine orange staining methods described by Bishop (Bishop and Smiles, 1957; Bishop and Austin, 1957) and Eweis and Schirren (1972), respectively, for use with ejaculated rat sperm (Goeden and Zenick, 1987). Acridine orange has been used to distinguish between living and dead cells since 1940.

This staining technique has been employed to assess spermatozoal viability in a number of species, including goat, mouse, and human. When examined under ultraviolet (UV) illumination, live sperm produce a red fluorescence when acridine orange solution is added, while nonviable sperm appear green. When rat sperm are incubated with a low concentration of acridine orange and examined under UV, the acrosome exhibits a red fluorescence similar to that shown by lysosomes in other cells. Under these conditions, the acrosome appears as a crescent-shaped image, extending over the nucleus. The nuclei not covered by an acrosome appear bright apple green.

D. In Vitro Measures of Sperm Function

The use of *in vitro* models for fertilization as screens for spermatotoxicity has been proposed. Evaluation of the influence of a xenobiotic on spermatozoal integrity and fertilizing ability has generally been conducted on vas deferens or epididymal samples incubated *in vitro* in the presence of the test agent. However, present *in vitro* techniques do not allow for the consideration of the dynamic and selective interaction between the male gamete and the female reproductive tract (Overstreet and Katz, 1977). The processes of sperm capacitation and fertilization are more efficient *in vivo,* in that the female reproductive tract plays an active role in sperm selection and transport and number capacitated.

Observations made employing *in vitro* techniques must be tested under more physiological conditions to determine if mechanisms discovered with *in vitro* experiments actually operate *in vivo.* In a recent study (Goeden and Zenick, 1987), we compared the spermatotoxicity of ethanol following *in vitro* or *in utero* incubation. Conflicting results were obtained. Rat spermatozoa *in utero* were not susceptible to the toxic effects of ethanol that were evident upon *in vitro* incubation. Many chemicals are capable of entering female reproductive tract fluids, thereby creating a potentially spermatotoxic environment. A dynamic, supportive environment within the tract may restrict the adverse effects. The inability of *in vitro* fertilization techniques to duplicate this component limits their utility in male reproductive risk assessment.

E. Evaluation of the Sensitivity of Various Indices of Spermatotoxicity and Their Relationship to Fertility

The relative sensitivity of a number of indices of spermatogenic function in the rat is indicated in Table I. These data, for the most part, are in close agreement with values reported elsewhere in the literature (i.e., Blazak *et al.*, 1985).

These data reflect the percent difference that can be detected for a change in a given measure as a function of sample size. Although the sample size recommended by EPA is 20 males per dose group, sample sizes in most studies fall well below this number (in fact, only 10 males per dose group are required in the FDA Segment 1 test). A common statistical error is to mate each male to two females and then treat the female as the unit of analysis. Since the male was the treated entity, he must constitute the unit for statistical analysis. Use of the females provides an inflated sample size that is inappropriate, since the pregnancies generated by a single male are not independent events.

Based on the data in Table I, testis weight would appear to be a highly sensitive measure. However, as discussed previously, it may be a relatively insensitive marker of gonadal toxicity. Of the spermatogenic indices listed, percent normal morphology has the least variation and has been shown to be indicative of toxicity with some agents in the absence of changes in other measures. Sperm morphology has also been shown to be a highly stable and sensitive parameter in humans (Wyrobek, 1983b).

Sperm counts derived from testicular samples appear to provide a somewhat more sensitive measure than counts based upon epididymal samples. This is, in part, attributed to the fact that the cauda epididymis of most test species contains large sperm reserves that interfere with the detection of moderate changes in sperm production (Hurtt and Zenick, 1986). However, if an agent has the potential to produce posttesticular effects or alter functional sperm characteristics (e.g., morphology, motility), vas deferens, epididymal, or ejaculated sperm evaluations must be conducted (e.g., methyl chloride; Working *et al.*, 1985a,b).

The data on motility presented in Table I were derived from videomicrographic techniques employed in our laboratory and from Blazak *et al.* (1985). This is the only measure in which the coefficients of variation are markedly different. However, these techniques have yet to be standardized across laboratories. Appropriate comparisons must await such standardization.

Limited data are available that have examined the independent and interdependent relationships between these various spermatogenic indices. Recent work has shown that spermatozoa with abnormal morphology exhibit poorer motility than normal sperm from the same ejaculate, whether the sample is from a fertile or infertile donor (Overstreet *et al.*, 1981; Katz *et al.*, 1982). However, the "normal-shaped" sperm from infertile men are also less likely to be motile and may swim more slowly than similarly shaped sperm from the semen of fertile men. These findings suggest that normal sperm from these two populations differ on physiological parameters not reflected in the standard evaluation.

Given that fertility tests are relatively insensitive, reproductive risk assessment may benefit from the inclusion of data on additional variables such as sperm measures. However, the validity of this position has not been thoroughly tested. Rarely do investigators utilize a sufficient array of end points to allow for a comparison of their relative sensitivities. Some support for using a more extensive protocol is provided by examining recent data on methyl chloride (Working *et al.*, 1985a,b) and ethylene glycol monomethyl ether (Chapin *et al.*, 1985a,b). Both of these investigations incorporated high-quality histopathology (including cell staging; Leblond and Clermont, 1952), sperm measures, and fertility assessments. The results of these studies are summarized in Tables II and III. For both agents, histopathological lesions and sperm alterations were seen at either lower doses and/or earlier in time than were effects on fertility or fetal outcomes. Similar relationships can be observed in reviewing the effects of dibromochloropropane (DBCP) in rabbits (Foote *et al.*, 1986a,b).

Aside from the issue of relative sensitivity of these various measures, the *quantitative relationship* of these indices to fertility is not well understood. Certain qualitative and quantitative standards must be met to insure fertility, but the lower limits of acceptability have not been delineated. For example, sperm counts in men that are lower than 20 million/ml have been traditionally taken as a sign of potential subfertility. However, there are men

TABLE II

Effects of Methyl Chloride Exposure on Various Endpoints of Male Reproductive Toxicity[a]

Endpoint	Dose (ppm)[b] 1000	3000
Fertility	—	D (weeks 2,3)
Fetal loss		
Preimplantation	I (week 3)	I (weeks 2–4,6,8)
Postimplantation	—	I (week 1)
Histopathology		
Epididymis	—	+ (week 2)
Testis	—	+ (weeks 1–8)
Spermatid counts	—	D (weeks 2–8)
Vas deferens sperm measures		
Sperm count	—	D (weeks 3–8)
Percent motile	—	D (weeks 1–8)
Percent abnormal forms	—	I (weeks 1–3)
Number of intact sperm	D (week 3)	D (weeks 2–8)

[a]Modified from Working *et al.* (1985a,b). Key: +, effect observed; –, no effect; D, decrease; I, increase.

[b]Five-day inhalation exposure.

TABLE III

Effects of Ethylene Gylcol Monomethyl Ether on Various Endpoints of Male Reproductive Toxicity[a]

Endpoint	Dose (mg/kg)[b] 50	100	200
Fertility	—	D (week 5)	D (weeks 4–16)
Fetal loss			
Preimplantation	—	I (week 5)	I (weeks 3–16)
Live fetuses	—	D (week 5)	D (weeks 4–16)
Resorptions	—	—	I (weeks 5–6)
Histopathology—testis[c]	+ (weeks 4–7)	+ (weeks 1–8)	+ (weeks 1–8)
Accessory-organ weight	—	—	—
Epididymal sperm measures			
Sperm count	D (week 5)	D (weeks 2–16)	D (weeks 2–16)
Percent motile	—	D (weeks 4–7)	D (weeks 3–16)
Percent abnormal forms	D (week 5)	D (weeks 5–16)	D (weeks 3–16)
Percent headless forms	—	I (weeks 5–?)	I (weeks 5–?)

[a]Modified from Chapin *et al.* (1985a,b). Key: +, effect observed; –, no effect; D, decrease; I, increase.

[b]Five-day oral exposure.

[c]Assessed for only 8 weeks.

whose sperm counts fall below this level who have fathered children, while there are others whose sperm counts exceeds the normal average (80–120 million/ml) but who are infertile. Thus, the distributions of sperm counts for fertile and infertile men overlap (Meistrich and Brown, 1983). It is likely that the standards for fertility are more stringent for humans than for traditional test species. Additional animal and human research is needed to better understand the biological consequences of spermatogenic alterations. At this stage, spermatotoxicity observed in laboratory studies must be of concern in the generalization and extrapolation of the risk to humans.

IV. New Approaches to Improving Quantitative Male Reproductive Risk Assessment

Increasing concern as to the sensitivity of the male reproductive system to chemical insult has been paralleled by reevaluation of current approaches and end points and their utility in reproductive risk assessment. As noted earlier, endpoints usually evaluated include histopathological analyses of reproductive tissue, breeding success of the male and subsequent fetal outcomes, and, with increasing frequency, semen evaluations. The combined application of these strategies provides a well-integrated approach to the first step in risk assessment, namely, hazard identification (National Academy of Science, 1983).

However, fertility assessment and sperm measures as currently evaluated have questionable sensitivity for the next phase in risk assessment, namely, dose-response quantification. Sperm production in most test species far exceeds the number required for fertility. Thus, substantial gonadal toxicity may occur before fertility is compromised. Moreover, the existence of large epididymal sperm reserves in these species may mask the detection of spermatotoxicity at low doses and/or result in an underestimation of the extent of actual impairment. Estimates of allowable human exposure levels based on such data may be inflated.

The ideal situation would be to employ a species (or strain within a species) wherein sperm production rates and subsequent sperm reserves are comparable to humans. To date, no test species meets such a criterion. However, it may be possible to develop strategies that reduce the large sperm reserves of the epididymides. Such manipulations might at least produce a functional analog to humans in terms of the number of sperm available for ejaculation. Reduction of these reserves could also facilitate the detection of spermatotoxicity at lower doses and earlier in time.

This hypothesis has been tested in a recent study (Hurtt and Zenick, 1986), wherein an active schedule of copulation was employed in rats to reduce epididymal reserves. The detection of spermatotoxicity in this animal relative to its nonmated counterpart was then compared following exposure to ethylene glycol monoethyl ether (ethoxyethanol, EE), a well-documented male reproductive toxicant (Oudiz *et al.*, 1984). Adult male Long–Evans hooded rats were assigned to a "mate" or "nonmate" condition, with the former mated every other day (3-h session) for 2 weeks prior to and then throughout the experiment. The latter were not mated. After 2 weeks, males from each group were randomly assigned to receive either 0, 150, or 300 mg/kg (p.o.) of EE, 5 days per week for 6 weeks. Males were then sacrificed and organ weights, testicular sperm counts, and cauda epididymal sperm counts and sperm morphology were evaluated.

The EE produced a significant reduction in testicular weight and spermatid counts in mated and nonmated animals receiving 300 mg/kg. Significant decreases were also noted in epididymal sperm count and morphology. However, these effects were seen in the nonmated animals only at 300 mg/kg, whereas significant reductions in both epididymal measures were

TABLE IV

Epididymal Sperm Count Following 6 Weeks of Ethoxyethanol Treatment[a]

	Dose (mg/kg, p.o.)		
Mate condition	0	150	300
Mate	519 ± 92^{c} (10)	$336 \pm 90^{d*}$ (12)	$219 \pm 95^{e*}$ (11)
Nonmate	757 ± 94^{b} (12)	722 ± 95^{b} (11)	$545 \pm 133^{c*}$ (13)

[a]Data are expressed as mean ± SD; number of animals evaluated in parentheses; sperm count expressed per gram epididymis $\times 10^6$.

[b,c,d,e]Means having same superscript are similar ($p > 0.05$; Kirk trend analysis).

*Significant at $p \leq 0.05$ compared to respective control; Duncan's multiple-range test.

also obtained at 150 mg/kg in males mated bidaily (Tables IV and V). Even when sperm count and sperm morphology were expressed as a percent of the respective control for the two mating conditions, the degree of decline was greater in the mated males at each dose for both indices (can be derived from Tables IV and V).

The approach utilized in this study was noninvasive and relatively simple to apply. It may be one of several strategies considered in attempting to develop better animal models to characterize low-dose effects that may be associated with a previously identified reproductive toxicant. The import of such models lies not in initial hazard identification, but rather in subsequent quantitative dose–response assessment (National Academy of Science, 1983).

TABLE V

Percent Normal Morphology Following 6 Weeks of Ethoxyethanol Treatment

	Dose (mg/kg, p.o.)		
Mate condition	0	150	300
Mate	98.0 ± 1.0^{b} (10)	$93.0 \pm 3.0^{c*}$ (12)	$82.0 \pm 4.0^{e*}$ (11)
Nonmate	98.0 ± 1.0^{b} (12)	98.0 ± 1.0^{b} (11)	$88.0 \pm 4.0^{d*}$ (13)

[a]Data are presented as mean ± SD; number of animals evaluated in parentheses.

[b,c,d,e]Means having same superscript are similar ($p > 0.05$; Kirk trend analysis).

*Significant at $p \leq 0.05$ compared to respective control; Duncan's multiple-range test.

References

Amann, R. P. (1981). A critical review of methods for evaluation of spermatogenesis from seminal characteristics. *J. Androl.* **2,** 37–58.

Bishop, M. W. H., and Austin, C. R. (1957). Mammalian spermatozoa. *Endeavor* **16,** 137–150.

Bishop, M. W. H., and Smiles, J. (1957). Induced fluorescence in mammilian gametes with acridine orange. *Nature (London)* **179,** 307–308.

Blazak, W. F., Ernst, T. L., and Stewart, B. E. (1985). Potential indicators of reproductive toxicity: Testicular sperm production and epididymal sperm number, transit time, and motility in Fischer 344 rats. *Fundam. Appl. Toxicol.* **5,** 1097–1103.

Chapin, R. E., Dutton, S. L., Ross, M. D., and Lamb, J. C. IV (1985a). Effects of ethylene glycol monomethyl ether (EGME) on mating performance and epididymal sperm parameters in F344 rats *Fundam. Appl. Toxicol.* **5,** 182–189.

Chapin, R. E., Dutton, S. L., Poss, M. D., Swaisgood, R. R., and Lamb, J. C., IV. (1985b). The recovery of the testes over eight weeks after short-term dosing with ethylene glycol monomethyl ether: histology, cell-specific enzymes and rete testis fluid proteins. *Fundam. Appl. Toxicol.* **5,** 515–525.

Chester, R. V., and Zucker, I. (1970). Influence of male copulatory behavior on sperm transport, pregnancy and pseudopregnancy in female rats. *Physiol. Behav.* **5,** 35–43.

Cullen, M. R., Kayne, R. D., and Robins, J. M. (1984). Endocrine and reproductive dysfunction in men associated with occupational inorganic lead intoxication. *Arch. Environ. Health* **39,** 431–440.

de Boer, P., van der Hoeven, F. A., and Chardon, J. A. P. (1976). The production, morphology, and karyotypes and transport of spermatozoa from tertiary trisomic mice and the consequence for egg fertilization. *J. Reprod. Fertil.* **48,** 249–256.

Eweiss, A., and Schirren, C. (1972). Experimental fluorescence microscopy of the so-called vital fluorochroming of human spermatozoa. *Andrologie* **4,** 129–138.

Foote, R. H., Schermerhorn, E. C., and Simkin, M. E. (1986a). Measurement of sperm quality, fertility, and reproductive hormones to assess dibromochloropropane (DBCP) effects in live rabbits. *Fundam. Appl. Toxicol.* **6,** 628–637.

Foote, R. H., Berndston, W. E., and Rounsaville, R. R. (1986b). Use of quantitative testicular histology to assess the effects of dibromochloropropane (DBCP) on reproduction in rats. *Fundam. Appl. Toxicol.* **6,** 638–647.

Goeden, H., and Zenick, H. (1987). Disposition of ethanol in blood and uterine fluid of estrous rats. *Biol. Reprod.,* in press.

Goy, R. W., Bischoff, K. B., Hogue, C. J., and Rosenberg, L. (1983). Other considerations: epidemiology, pharmacokinetics, and sexual behavior. *In* "Assessment of Reproductive and Teratogenic Hazards" (M. S. Christian, M. W. Galbraith, P. Voytek, and M. A. Mehlman, eds.), pp. 117–144. Princeton Scientific, Princeton, New Jersey.

Hurtt, M., and Zenick, H. (1986). Reduction of epididymal reserves enhances detection of ethoxyethanol-induced spermatotoxicity. *Fund. Appl. Toxicol.* **7,** 348–353.

Jeulin, C., Soumah, A., and Jouannet, P. (1985). Morphological factors influencing the penetration of human sperm into the cervical mucus *in vitro. Int. J. Androl.* **8,** 215–233.

Johnson, L., Petty, C. S., and Neaves, W. B. (1980). A comparative study of sperm production and testicular composition in humans and rats. *Biol. Reprod.* **22,** 1233–1243.

Katz, D. F., and Overstreet, J. W. (1981). Sperm motility assessment by videomicrography. *Fertil. Steril.* **35,** 188–193.

Katz, D. F., Diel, L., and Overstreet, J. W. (1982) Differences in the movement of morphologically normal and abnormal human seminal spermatozoa. *Biol. Reprod.* **26,** 566–570.

Lancranjan, I., Popescu, H. I., Gavanescu, O., Klepsch, I., and Serbanescu, M. (1975). Reproductive ability of workmen occupationally exposed to lead. *Arch. Environ. Health* **30,** 396–401.

Leblond, C. P., and Clermont, Y. (1952). Definition of the stages of the seminiferous epithelium in the rat. Part 1. Spermatogenesis and sperm maturation. *Ann. N.Y. Acad. Sci.* **55,** 548–573.

Meistrich, M. L. (1982) Quantitative correlation between testicular stem cell survival sperm production and fertility in the mouse after treatment with different cytotoxic agents. *J. Androl.* **3,** 58–68.

Meistrich, M. L., and Brown, C. C. (1983). Estimation of the increased risk of human infertility from alterations in semen characteristics. *Fertil. Steril.* **40,** 220–230.

National Academy of Science (1983). "Risk Assessment in the Federal Government Managing the Process." National Academy Press, Washington, D.C.

Nestor, A., and Handel, M. A. (1984). The transport of morphologically abnormal sperm in the female reproductive tract of mice. *Gamete Res.* **10,** 119–125.

Nelson, J., and Zenick, H. (1986). The effect of trichloroethylene on male sexual behavior: Possible opiod role. *Neurobehav. Toxicol. Teratol.* **8,** 441–445.

Niwa, K., and Chang, M. C. (1974). Various conditions for fertilization of rat eggs *in vitro. Biol. Reprod.* **11,** 463–469.

Oudiz, D., Zenick, H., McGinnis, P., and Niewenhuis, R. J. (1984). Male reproductive toxicity and recovery associated with acute and subchronic ethoxyethanol exposure. *J. Toxicol. Environ. Health* **13,** 743–755.

Overstreet, J. W., and Katz, D. F. (1977). Sperm transport and selection in the female genital tract. *In* "Development in Mammals," (M. Johnson, ed.), Vol. 2. North Holland, New York.

Overstreet, J. W., Price, M. J., Blazak, W. F., Lewis, E. L., and Katz, D. F. (1981). Simultaneous assessment of human sperm motility and morphology by videomicrography. *J. Urol.* **126,** 357–360.

Paz-Frenkel, G., Kaplan, R., Yedwab, G., Homonna, Z., and Kracier, P. F. (1978). The effects of caffeine on rat spermatozoa motility, metabolism, and fertilizing capacity. *Int. J. Androl.* **1,** 145–152.

Ratnasooriya, W. D., Gilmore, D. P., and Wadsworth, R. D. (1980). Antifertility effect of sympathomimetic drugs on male rats when applied locally to the vas deferens. *J. Reprod. Fertil.* **58,** 19–25.

Redi, C. A., Garagna, S., Pellicciari, C., Manfredi Romanini, M. G., Cappana, E., Winking, H., and Gropp, A. (1984). Spermatozoa of chromosomally heterozygous mice and their fate in the male and female genitals tracts. *Gamete Res.* **9,** 273–286.

Robaire, B., Smith, S., and Hales, B. F. (1984). Suppression of spermatogenesis by testosterone in adult male rats: effect on fertility, pregnancy outcome and progeny. *Biol. Reprod.* **31,** 221–230.

Robb, G. W., Amann, R. P., and Killian, G. J. (1978). Daily sperm production and epididymal sperm reserves of pubertal and adult rats. *J. Reprod. Fertil.* **54,** 103–107.

Rubin, H. B., and Henson, D. E. (1979). Effects of drugs on male sexual function. *In* "Advances in Behavioral Pharmacology." Vol. 2. pp. 65–86. Academic Press, New York.

Schwetz, B. A., Rao, K. S., and Park, C. N. (1980). Insensitivity of tests for reproductive problems. *J. Environ. Pathol. Toxicol.* **3,** 81–98.

Sokol, R. Z., Madding, C. E., and Swerdloff, R. S. (1985). Lead toxicity and the hypothalamic-pituitary axis. *Biol. Reprod.* **33,** 722–728.

Tepe, S., and Zenick, H., (1984). The effects of carbon disulfide on the reproductive system of the rat. *Toxicology* **31,** 47–56.

Tessler, S., and Olds-Clarke, P. (1985). Linear and nonlinear mouse sperm motility patterns. *J. Androl* **6,** 35–44.

Waller, D. P., Killinger, J. M., and Zaneveld, L. J. D. (1985). Physiology and Toxicology of the male reproductive tract. *In* "Endocrine Toxicology" (J. A. Thomas, ed.), pp. 270–333. Raven, New York.

Working, P. K., Bus, J. S., and Hamm, T. E., Jr. (1985a). Reproductive effects of inhaled methyl chloride in the male Fischer 344 rat. I. Mating performance and dominant lethal assay. *Toxicol. Appl. Pharmacol.* **77,** 133–143.

Working, P. K., Bus, J. S., and Hamm, T. E., Jr. (1985b). Reproductive effects of inhaled methyl chloride in the male Fischer 344 rat. II. Spermatogonial toxicity and sperm quality. *Toxicol. Appl. Pharmacol.* **77,** 144–157.

Wyrobek, A. J., Gordon, L. A., Burkhart, J. G., Francis, M. W., Kapp, R. W., Jr., Letz, G., Malling, H. V., Topham, J. C., and Whorton, M. D. (1983a). An evaluation of the mouse sperm morphology test and other sperm tests in nonhuman mammals. *Mutat. Res.* **115,** 1–72.

Wyrobek, A. J., Gordon, L. A., Burkhart, J. G., Francis, M. W., Kapp, R. W., Letz, G., Malling, H. V., Topham, J. C., and Whorton, M. D. (1983b). An evaluation of humans perm as indicators of chemically-induced alterations of spermatogenic function. *Mutat. Res.* **115,** 73–148.

Wyrobek, A. J., Watchmaker, G., and Gordon, L. (1984). An evaluation of sperm tests as indicators of germ-cell damage in men exposed to chemical or physical agents. *In* Reproduction, The New Frontier in Occupational and Environmental Research," Program in Clinical Biological Research (J. E. Lockey, G. K. Lemasters, and W. R. Keye Jr., eds.), Vol. 160, pp. 385–406. Alan R. Liss, New York.

Zenick, H., Blackburn, K., Hope, E., Oudiz, D., and Goeden, H. (1984a). Evaluating male reproductive toxicity in rodents: A new animal model. *Teratogen. Carcinogen. Mutagen.* **4,** 109–128.

Zenick, H., Blackburn, K., Hope, E., and Baldwin, D. J. (1984b). An assessment of the copulatory, endocrinologic, and spermatotoxic effects of carbon disulfide exposure in rats. *Toxicol. Appl. Pharmacol.* **73,** 275–283.

Zenick, H., Blackburn, K., Hope, E., Smith, K., and Richdale, N. (1984c). Effects of trichloroethylene exposure on male reproductive function. *Toxicology* **31,** 247–250.

Zenick, H., Hope, E., and Smith, K. (1986). Reproductive toxicity associated with acrylamide treatment in male and female rats. *J. Toxicol. Environ. Health* **17,** 457–472.

Chapter 9

Genetic Toxicology of Mammalian Male Germ Cells

MICHAEL D. SHELBY

Cellular and Genetic Toxicology Branch
National Institute of Environmental Health Sciences
Research Triangle Park, North Carolina 27709

I. Introduction

A. Genetic Toxicology

The science of toxicology encompasses a wide range of disciplines, all concerned with the harmful health effects that result from exposure to chemical or physical agents. Except for one, all these disciplines concern themselves with effects on the exposed individual. This one exception is the genetic toxicology of germ cells, wherein the concern is not for the individual exposed, but rather for the offspring of the exposed individual and how the offspring's health may be affected detrimentally by damage induced in the gametes of the exposed parent.

In order to determine whether an agent is capable of inducing heritable damage, it is necessary to conduct tests using mammalian germ cells. The general scheme for all such tests is to treat one or both parents with the agent under study and then to mate the parents and assess the offspring for evi-

PHYSIOLOGY AND TOXICOLOGY
OF MALE REPRODUCTION

dence of induced, inherited genetic change. Consistent with the theme of this book, discussions in this chapter will be limited to studies in which the male is the treated parent. Further, discussion will be limited to studies in mice. Although any laboratory mammal could be used for germ cell studies, and most have been to some extent, the mouse is by far the most commonly used. This is primarily because of its small size, ease in breeding, and the extensive body of knowledge of mouse genetics.

The genetic material of mammals, as with all eukaryotes, is organized in chromosomes that consist of DNA and associated proteins. The types of genetic damage of concern in germ cell genetic toxicology can be categorized into three groups that parallel the organization of the genetic material. At the DNA level, changes in the sequence of nucleotide bases may occur as the result of addition or deletion of bases or of the substitution of one base for another during DNA replication or repair. Such changes normally affect the function of a single gene and are referred to as gene (or point) mutations.

At the level of the chromosome structure, breaks may occur and result in chromosome fragments, and, if fragments rejoin, chromosome rearrangements. This type of damage is referred to as chromosomal aberrations.

The third general type of genetic damage involves the number of chromosomes. As any species is characterized by having its genetic material organized into a defined number of chromosomes, any departure from that number affects the genetic integrity of the individual. When a gamete contains too many or too few chromosomes, the product of its fertilization will likewise bear an abnormal number of chromosomes. This departure from the normal number of chromosomes is referred to as aneuploidy. When the chromosome number is increased above the normal by multiple complete sets of chromosomes, the condition is referred to as polyploidy.

Tests that involve the treatment and breeding of mice are available for detecting all three types of genetic damage discussed above (see Table I). In addition, studies can be conducted in the treated males to determine whether a test agent actually reaches the germ cells and whether the agent is interacting with the DNA in such a way as to elicit DNA repair.

Following a brief discussion of the basis for concern that human exposure to germ-cell mutagens poses a health risk to future generations, assay systems for detecting various types of genetic damage in mice will be discussed individually.

B. Basis for Concern

Because the genetic condition of the germ cells is of no consequence to one's personal health, it is worthwhile to briefly consider the health concern on which germ-cell genetic toxicology is based. Simply stated, the concern is

TABLE I

Summary of Characteristics of Germ-Cell Genetic Toxicity Tests

Test	Phenomenon observed	Genetic end point detected	Heritability
Dominant lethal	Deciduomata	Chromosome abberations	No
Heritable translocation	Fertility/sterility	Reciprocal translocations	Yes
MSLT	External features	Gene mutation/small deletion	Yes
BSLT	Electrophoretic pattern	Gene mutation/small deletion	Yes
Recessive lethal	Uterine death	Gene mutation	Yes
Skeletal	Abnormal skeleton	Dominant mutations	Yes
Cataract	Cataracts	Dominant mutations	Yes
Cytogenetic analyses	Aberrations	Chromosome aberrations	No
Aneuploidy	Abnormal chromosome number	Missegregation	Yes
Gonadal exposure	Radioactivity	DNA repair or DNA binding	No

that mutations, either spontaneous or induced, in the germ cells may result in offspring that suffer from genetically based disease. Such diseases are occasionally treatable, never curable, and may be passed on to subsequent generations.

Hermann J. Muller, Nobel laureate, was the first to demonstrate the induction of germ-cell mutations by an environmental agent. In 1929, he showed that ionizing radiation induced mutations in the germ cells of the fruit fly *Drosophila melanogaster.* Subsequently, he became a leader in the movement to protect human genetic welfare from the detrimental effects of radiation.

In the 1940s, the germ-cell mutagenicity of chemicals was clearly demonstrated by Charlotte Auerbach, again in the fruit fly. As the number of known chemical mutagens grew, so did the concern that exposure to environmental agents might impact on the health of the human population through the induction of germ-cell mutations and result in an increased incidence of genetic disease.

Germ-cell genetic toxicology occupies a unique position in the overall field of toxicology because, to date, no instance of genetic disease resulting from induced mutation has been demonstrated in humans. Nonetheless, the facts remain that chemicals can be mutagenic to mammalian germ cells, that humans are exposed to mutagenic chemicals, and that mutations can result in genetic disease.

Because biological evolution is based on genetic variability and mutations underlie that variability, it has been suggested that mutations in human populations may be beneficial. This notion, however, is refuted by both empirical evidence and theoretical considerations. Based on current knowledge of the incidence and effect of genetic disease in the human population, biological evolution, and population genetics, there can be little doubt that the health of future generations will be served by efforts to minimize human exposure to germ-cell mutagens.

Because it is unacceptable to conduct tests on humans, and submammalian test systems are poor models for studying germ-cell mutations, the mouse provides the preferred models for testing and studying germ-cell mutagens. As discussed later, test systems of variable utility are available to investigate all general classes of genetic damage.

Depending on the type and amount of data available, two types of human genetic risk estimates can be made, absolute or relative. An absolute risk estimate predicts the number of individuals who will suffer from genetic disease as the result of a population's exposure to a germ cell mutagen. The relative risk estimate predicts genetic disease incidence as a fraction of the existing or "normal" incidence of genetic disease. If the existing incidence is known, either estimate could be converted to the other.

Unfortunately, the human genetic disease burden is poorly understood. The most frequently quoted figure is 10% for the portion of disease that is genetically based. If all disease states for which there is a subtle genetic component are considered, that figure will be higher. Similarly, it would be much lower if only those diseases are considered that exhibit strict Mendelian patterns of inheritance, that is, clearly demonstrable dominant and recessive traits. A further complication in the consideration of human genetic disease is posed by the uncertainty of how much genetic disease results from new mutations versus the amount attributable to detrimental genes that already exist in the human gene pool.

Despite limited knowledge of the origin and true incidence of human genetic disease, the limitations of available rodent systems for studying induced mutations, and the numerous assumptions required in calculating estimates of genetic risk, the fact remains that most of what we know about genetics, mutagenesis, and genetic disease points toward a tangible health risk associated with exposure to germ cell mutagens. For this reason, it is important to determine if a risk exists and, if so, to minimize this risk of genetic disorders to yet unborn generations.

C. General Testing Considerations

In conducting a mammalian germ-cell mutagenicity testing program, two issues that must be dealt with, regardless of the tests to be conducted, are which chemicals to test and by what route of exposure. Both choices should be influenced by the physical and chemical characteristics of the chemicals being considered and by the nature, route, and extent of human exposures to those chemicals. However, some general guidelines can be offered.

Some agents are intrinsically mutagenic either directly or following metabolic modification. It is only intrinsically mutagenic chemicals that are potential germ-cell mutagens (see Table II). Because the intrinsic mutagenicity of a chemical can be determined using *in vitro* or *in vivo* tests that require less resources than germ-cell mutagenicity assays, germ-cell tests should be conducted only on chemicals with demonstrated mutagenicity. Evidence that a chemical is mutagenic in somatic cells of intact rodents provides a stronger basis for its selection for germ-cell tests than does evidence derived only from *in vitro* assays. All germ-cell mutagens defined to date also induce genetic damage in somatic cells, although not all somatic cell mutagens are germ-cell mutagens. It is also advantageous, before conducting the more expensive germ-cell tests, to have evidence that the test agent or a metabolite reaches the male germ cells; this evidence can be obtained from investigations of chemical distribution, reproduction, or other similar studies. Fi-

TABLE II

Examples of Chemicals Producing Positive Results (+) in Some Mouse Germ-Cell Mutagenicity Tests[a]

	DL	HTT	BSLT	MSLT
Cyclophosphamide	+	+		+
Ethyl methanesulfonate	+	+		+
Ethylnitrosourea	+	+	+	+
Procarbazine	+	+	+	+
Trethylenemelamine	+	+	+	+

[a]Key: DL, dominant lethal test; HTT, heritable translocation test; BSLT, biochemical specific-locus test; MSLT, morphological specific-locus test; +, positive response detected.

nally, only those chemicals for which significant human exposure is known or anticipated justify the investment of testing resources.

Routes of exposure for germ-cell studies have included all those typical of rodent toxicology studies: inhalation, feed, water, gavage, intraperitoneal injection, and subcutaneous injection. Selection of exposure route typically is based on some compromise between an attempt to mimic human exposure, in terms of both route and concentration, and attempts to attain systemic exposure levels in the test animals that will ensure that even weak germ-cell mutagens will be detected. The latter consideration generally outweighs the former, so that potential germ-cell mutagens are not falsely labeled negative. Therefore, most studies have been conducted using near-lethal doses of test chemicals, administered acutely by routes that are likely to give high blood or tissue levels, such as, gavage, inhalation, or intraperitoneal injection.

II. Test Methods

A. Dominant Lethal Test

Damage may be induced in the genetic material of the developing male germ cell that does not affect the ability of a germ cell to fertilize but results in the death of the conceptus at one of the early cell divisions. Experimental evidence suggests that the genetic damage responsible for dominant lethality is chromosomal aberrations and that the death of the conceptus results from imbalances in the normal chromosomal complement following cell division (e.g., loss of fragments of chromosomes). It has also been suggested that, in addition to chromosomal aberrations, aneuploid gametes (having too few or too many chromosomes) will lead to dominant lethality and hence, constitute a second class of genetic damage that is detected by the dominant lethal assay.

By the nature of the event detected in this assay (i.e., death), it is clear that the test does not assess a biological endpoint that reflects a potential risk to future generations. Nonetheless, in efforts to detect and evaluate environmental factors that pose a genetic risk, the dominant lethal assay plays a key role in initial attempts to determine whether an agent reaches the germ cells and induces genetic damage (i.e., chromosomal aberrations). Although the induction of dominant lethality poses no risk to future generations, it provides clear evidence that the integrity of the germ-cell genetic material has been affected, and this is adequate basis for concern that other more serious, heritable effects may also be induced by the agent under study.

Any sexually reproducing organism could, theoretically, be used to conduct dominant lethal studies. However, in the field of genetic toxicology where the basis for concern is effects on human germ cells, laboratory mammals are the organisms of choice. In the past, rats were used extensively in these tests, but the mouse is now the most widely used animal, probably because of its small size and ease of breeding. It is also the animal used in nearly all other mammalian germ-cell mutagenicity studies and thereby allows a more direct comparison among tests.

Females may be used as the treated parent in dominant lethal studies, but because early embryonic survival may be affected by induced changes in female reproductive physiology and because of the limitations on the numbers of litters and offspring obtained from treated females, males are usually the parent treated in these studies.

Outbred, inbred, and hybrid mice have all been used in dominant lethal studies, and the choice is usually based on availability or the preference of a given laboratory. There are at least two factors that should be considered in this choice. First, a cross should be selected that yields a low and relatively stable frequency of spontaneous dominant lethal events. By minimizing variability in the background, data interpretation is simplified and greater confidence is imparted to test results. Second, the strain of female should be carefully considered, as it has been shown that the strain of untreated female can have a major effect on the observed frequency of dominant lethality.

Numerous factors influence the number of animals, both treated males and mated females, to be used in a dominant lethal test. These include the purpose of the test (e.g., regulatory requirements may be specific on this issue), variability in background frequencies and the statistical analyses employed, and number of doses and dosing regimen employed. In general, if one is to detect with 95% confidence a doubling of the spontaneous rate at a given dose and germ-cell stage, approximately 50 mated females are required. Depending on the mating scheme used, 20–50 treated males may be required.

The basic design of a dominant lethal test is treatment of the male parent with the agent under study, mating of treated and control males with untreated females, and analysis of uterine contents of females 14–16 days after mating. The incidence of living and dead implants is recorded and data are analyzed. Definitive dominant lethal effects are evidenced by an increase in dead implants and a decrease in live implants.

B. Heritable Translocation Test

The heritable translocation test (HTT) is designed to detect chromosomal damage in germ cells: specifically, reciprocal exchanges of chromosome segments between nonhomologous chromosomes. Germ cells bearing reciprocal translocations are normally capable of fertilization, and the resulting individual, having a complete complement of chromosomes, is viable. The primary biological effect in the translocation-carrying individual is reduced fertility or, in some cases, complete sterility. The effect on fertility results from disruption of an event unique to meiosis when homologous chromosomes pair in the first meiotic division. Because the two translocated chromosomes have no complete homolog, pairings involve more than two chromosomes and can result in the disruption of meiosis. When chromosomes are successfully distributed into the germ cells, theoretically half of the resulting gametes will bear a duplication of some chromosomal material (the distal material on the translocated chromosome) and a deletion of the chromosome segment that was exchanged with that distal segment (Fig. 1).

As in most of the tests of germ-cell mutagenicity, the induction of heritable translocations can be tested in either males or females, but for practical considerations presented earlier, the tests are routinely conducted using treated males. These males are mated to untreated females. F_1 females are not evaluated further, and F_1 males are then mated to identify semisterile or sterile males. Males that exhibit reduced fertility are then assessed cytogenetically to determine if the effect is indeed the result of a translocation. In semisterile males, preparation of testicular tissue to permit visualization of chromosomes in first division spermatocytes will reveal translocations as multivalent chromosome configurations. In fully sterile males where total disruption of meiosis precludes cytogenetic analysis of spermatocytes, it is necessary to evaluate metaphase preparations from somatic cells, such as kidney epithelium or lymphocytes. Due to the resources and expertise required to conduct the HTT and the limited number of laboratories that routinely conduct the test, a limited historical data base exists on the spontaneous incidence of heritable translocations in mice. Nonetheless, this rare event is estimated to occur spontaneously in about one in 5000 male gametes. Based on this figure, it is recommended that tests of chemicals for the

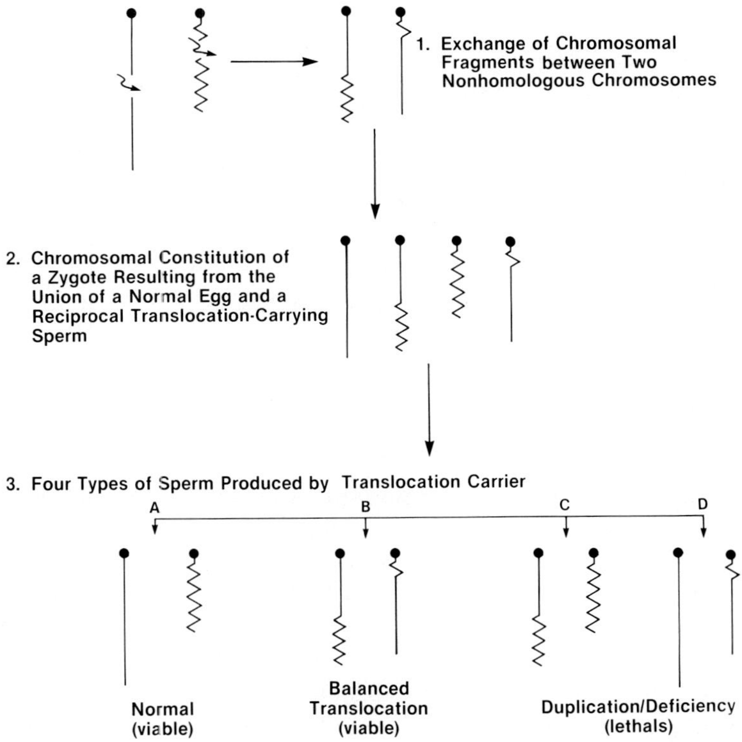

Fig. 1. Reciprocal translocations in postmeiotic sperm. An exchange of chromosomal material between two nonhomologous chromosomes is depicted along with the chromosomal constitution of the four sperm types produced by an F_1 male derived from the zygote produced by the translocation-bearing sperm.

induction of heritable translocations include the analysis of at least 500 F_1 males.

C. Morphological Specific Locus Test

The major concerns in genetic risk considerations have been gene mutations in gonial stem cells (Fig. 2). The stem cells are the origin of germ cells throughout the reproductive life of an organism, and their permanence provides the only germ-cell stage wherein genetic damage can accumulate through time and thereby pose an increasing risk of genetic damage to the progeny. Further, gene mutations are less likely than chromosomal damage to be eliminated by selection during the cell divisions of gametogenesis and are, therefore, more apt to exist in mature germ cells. As can be seen from Fig. 2, matings that occur in week 7 or later weeks following treatment of the

Day of Mating Following Treatment of Male Parent	Treatment 0 · 1 2 3 4 5 6 7 · 8 9 10 11 12 13 14 15 16 17 18 19 20 21 · Meiosis · 22 23 24 25 26 27 28 29 30 31 32 33 34 · 35 36 37 38 39 40 41 42 · 43——Later
Stage of Development when Treatment Occurred	Mature Sperm · Spermatids · Spermatocytes · Differentiated Spermatogonia · Spermatogonial Stem Cells

Fig. 2. Relationship between day of mating posttreatment and stage of sperm at time of treatment.

male parent utilize sperm that were in the spermatogonial stem-cell stage at the time of treatment. By doing serial matings after treatment of the male, one can determine the germ-cell stage(s) sensitive to the effect of the test agent.

The most extensively used method for studies of mutations induced at specific loci in mammalian germ cells is the mouse morphological specific-locus test for visible markers (MSLT), developed by W. L. Russell in the early 1950s. This test was developed to assess the germ-cell mutagenicity of ionizing radiation and provided the bulk of data used in radiation genetic risk analysis in past years. More recently, the test has been used to study the germ-cell mutagenicity of chemical agents.

The basic concept of the MSLT is relatively simple: it involves seven genes (or loci) that affect morphological characteristics of the mouse, namely, hair color, eye color, and ear shape. In the standard test, males homozygous for the wild-type alleles of these seven genes are treated with the test agent and then mated to untreated females homozygous for recessive alleles at the same seven loci. Normal male gametes will result in progeny that are heterozygous at these seven loci and will have the same wild-type appearance as their sires. If treatment of the males has induced mutations that affect the expression of any of the seven loci, sperm containing such mutations will give rise to offspring that are homozygous recessive at one (or more) loci and will express the phenotype of the recessive allele (Fig. 3). At 2–4 weeks of age, progeny are examined for variant appearance; normal progeny are discarded, and variants are kept to confirm, by breeding tests, that the variant phenotype is genetically based—that is, is transmitted to the offspring.

The types of genetic damage recovered in this test system include intragenic changes such as base-pair substitutions, small additions or deletions of genetic material, and deletions of larger segments of the DNA that spans the specific loci studied and adjacent DNA that may include other marker genes. Using recombinant DNA techniques, it has recently been demonstrated that among spontaneous mutants at the dilute locus, some mutations result from the insertion of a viral DNA sequence within the locus. Preliminary evidence indicates that the same may be true for the agouti locus. Studies have been undertaken to determine whether insertion elements are associated with induced mutations at these loci.

Specific locus tests for visible markers are not restricted to the seven loci used by Russell, and other laboratories have constructed mouse strains to assess mutations at other loci. It must be noted that the number of loci that can be utilized in an assay for visible markers is restricted by their availability and their compatibility within such an assay.

Because of the long history of the MSLT, an immense historical data base has accumulated on the frequency of spontaneous mutations at the loci studied. This data base offers the assay two great advantages. By virtue of the evidence for low variability across time, it mitigates the need for large concurrent controls, and it provides a sound basis for the development of sensitive statistical analyses of test data.

As presently conducted, approximately 50 males per dose group are mated with two females per week until an adequate number of progeny has been obtained to reach a conclusion on the mutagenicity of the test agent in spermatogonia. By current criteria, a positive effect can be detected in less than 1000 offspring if any mutants are recovered and a negative conclusion can be reached with less than 12,000 offspring.

Despite the widely held belief that specific-locus tests are prohibitively expensive, a single-dose experiment testing male spermatogonia can be conducted for a fraction of the cost of a rodent cancer study. Tests of additional

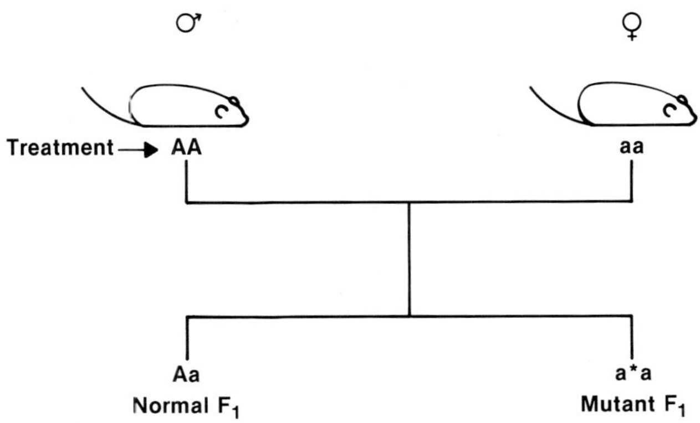

Fig. 3. Specific-locus test. The basis for the morphological specific locus test in which a homozygous wild-type male parent is treated with the test agent and then mated to a homozygous recessive female. Mutants are detected as offspring expressing a recessive trait; a*, mutant gene induced in male parent results in expression of recessive trait in offspring.

doses, other male germ-cell stages, and on treated females will of course drive this cost substantially higher, but for those situations where knowledge of germ-cell mutagenicity is needed to assess risk in the interest of human health, the cost of necessary studies cannot be considered prohibitive.

The loci studied in these mouse assays are not related to loci that affect murine or human health and are only presumed to be representative of other loci in the genome. They were chosen for convenience and provide a measure of the frequency with which mutations are induced by a test agent. This information must be considered in conjunction with multiple other factors and assumptions in the process of risk analysis.

D. Biochemical Specific-Locus Test

The biochemical specific-locus test (BSLT) is similar in many respects to the MSLT, and only the unique characteristics of the BSLT will be discussed here.

Two inbred strains of mice, C57BL/6 and DBA/2, are used in the BSLT and, as in the MSLT, the standard test involves treatment of males (usually DBA) and mating with untreated females of the other strain. The F_1 progeny are then analyzed for evidence of mutations.

Tissue samples, either blood from the orbital sinus or kidney tissue from heminephrectomy, are taken from each F_1 animal and subjected to either starch gel electrophoresis or isoelectric focusing.

In this test, mutations are detected that affect the electrophoretic characteristics of defined products of genes that have been mapped on mouse chromosomes (Fig. 4). Two features make the test particularly desirable for germ-cell mutagenesis studies. The test screens for mutations in a large number of known genes, thereby providing an efficiency that is not available in other gene mutation tests. At present, more than 30 genes are screened in each F_1 animal analyzed as compared to the seven screened in the MSLT. In addition, these genes are distributed throughout the genome, having been mapped to 15 different chromosomes ($n = 20$ in the mouse) and, theoretically, a much larger number of genes could be screened as technological developments make such screening feasible.

In addition, many of the gene products analyzed are similar to human gene products. This being so, more confidence may be placed in the extrapolation of mouse test results to potential genetic risk in humans. Furthermore, some of the gene products screened in mice are the same as those that can result in genetic disease in humans, again providing a closer link between rodent test results and possible human health effects. Mutations recovered in genes that are related to genetic disease in humans may also provide invaluable mouse models for human diseases (e.g., Cooley's anemia) and contribute to the further understanding and treatment of genetic diseases.

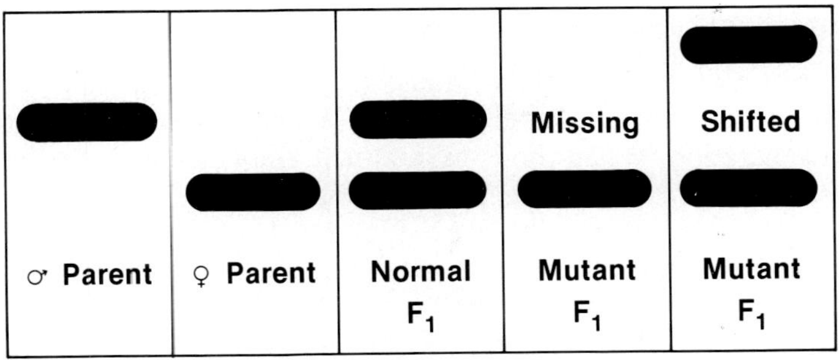

Fig. 4. Biochemical specific-locus test: electrophoretic patterns. Examples of the types of variant electrophoretic patterns detected in a protein with different electrophoretic mobility in the two parental strains.

The most readily identifiable limitations of this test system when compared with the MSLT are the equipment and expertise needed to conduct the surgery and electrophoresis procedures.

E. Recessive Lethal Test

The induction of recessive lethal mutations at unspecified (unknown) loci can be detected by treatment of a parental generation, production of F_1 and F_2 generations, and back-crossing of F_2 females to their sires. Pregnant females in the back-cross are sacrificed and uterine contents assessed for the evaluation in incidence of dead implants, which would indicate the F_1 male to be heterozygous for a recessive lethal mutation in either the gene or chromosomal level. Although this test would provide valuable data with regard to mutations with detrimental effects and would theoretically sample the entire genome, it has been considered too cumbersome for chemical mutagenicity studies.

Similarly, tests that employ chromosome inversions and screen for recessive lethal mutations in small segments of the genome have not been used extensively because of the expense, even though the results would be valuable.

F. Dominant Mutations

Dominant mutations, by definition, are expressed in the heterozygous state. Therefore, induced detrimental dominant mutations will affect the first generation following exposure to a germ-cell mutagen.

Two test systems have been developed to detect and quantify the induction of dominant mutations; one detects skeletal abnormalities and the other detects cataracts.

1. Dominant Skeletal Mutation Test

This test employs two inbred mouse strains. Typically, male parents are treated and mated to untreated females. The progeny are bred to produce a second generation. When the F_2 is obtained, the F_1 animals are killed and their skeletons are examined. When abnormalities are observed, the offspring (F_2) of variant animals are examined to determine if the skeletal trait was passed on as a dominant mutation. The production of an F_2 generation, although necessary to confirm the heritability of an abnormality, is expensive and time-consuming. For this reason, an abbreviated version of the test has been utilized. In this "sensitive indicator" system, a few characteristic skeletal abnormalities that have been demonstrated in earlier studies to result most frequently from dominant mutations are employed. The occurrence of these abnormalities in the F_1 is assumed to result from dominant mutations, and no F_2 generation is produced.

Calculations based on the frequency of induced mutations in this system, along with several assumptions including correction factor that extends the results to all major body systems, have been used to derive estimates of absolute risk from radiation exposure, that is, to estimate the number of individuals expected to be affected following parental exposure to ionizing radiation.

2. Dominant Cataract Mutation Test

In mice it is known that cataracts (clouding of the cornea) can be an inherited trait. A method has been developed wherein the progeny of mutagen treated parents are screened for the presence of cataracts. The F_1 animals are examined using a slit lamp after dilation of their pupils. In contrast to the skeletal system, progeny are not sacrificed for examination, which greatly simplifies the testing of variants for heritability. Only those animals exhibiting cataracts need to be bred to test heritability. Again, data suitable for calculating estimates of absolute risk are provided.

3. Dominant Morphological Mutants

A third and less well-defined method for detecting dominant mutations is incidental to all assays in which an F_1 generation is produced. It is a simple matter to observe all F_1 progeny for evidence of gross morphological variations and to breed morphologically variant animals to determine heritability of the variant traits.

Although dominant mutation systems test for mutations in genes for which neither the number nor location are known, they are thought to detect mutations at a large number of loci scattered throughout the genome.

To increase the efficiency of testing by obtaining as much information as possible from each F_1 animal derived from the treated parental generation, it is possible to combine a number of the assays described above. Ehling and his co-workers have reported results from tests that combine visible and biochemical endpoints with the cataract assay. In this system it is possible to screen an estimated 70 genetic loci in each F_1 animal, a substantial increase over the seven loci per animal in the visible specific-locus test or even the 30-plus loci in the biochemical specific-locus test.

G. Cytogenetic Tests

Cytogenetics methods offer means by which chromosomal damage induced in male germ cells can be detected and quantified. Cytogenetic analysis can be conducted only on cells that are undergoing nuclear division, because it is only during mitosis or meiosis that the chromosomes are in a physical state that allows microscopic examination. In the developing male germ cells, mitotically dividing differentiated spermatogonia and spermatocytes in the first or second meiotic division can be cytogenetically analyzed. Following the completion of meiosis, neither spermatids nor spermatozoa can be cytogenetically analyzed, but the sperm chromosomes can be examined in the zygote following fertilization. Prior to the first division of the zygote, the maternal and paternal pronuclei can be distinguished, and as their chromosomes decondense, they can be scored for structural or numerical anomalies.

Most chromosomal damage results in gross imbalances in the chromosomal constitution following cell division and is lethal to daughter cells. Hence, although cytogenetic analysis may provide irrefutable evidence of induced genetic damage in the germ cells, the majority of this damage does not pose a risk of genetic disorders to the liveborn offspring. The majority of chromosomally abnormal cells is eliminated during mitotic or meiotic divisions in spermatogenesis, or the conceptus dies at an early stage of development (e.g., as measured in the dominant lethal test). The two general categories of cytogenetic damage that can lead to genetically impaired offspring are inversions and translocations.

The lethal nature of most chromosomal aberrations results in the rapid loss of most cytogenetically abnormal cells and requires the investigator to schedule tests so that the cells to be scored are in their first mitotic division following treatment. For detection of damage induced in differentiated spermatogonia, animals are normally sacrificed 24 h following treatment.

H. Evidence for Gonadal Exposure

In addition to the direct measurements of induced genetic effects in the germ cells, it is informative and often desirable to obtain information on whether or not a test agent has reached the germ cells. Several methods have been used to determine if a chemical has reached this target tissue. The acquisition of such data has at least two purposes. If nothing is known about the germ-cell effects of a chemical, one may choose to conduct preliminary studies on the presence and/or effect in the germ cells. If the agent or its metabolite does not reach the germ cells, it will not induce mutations there. The second general purpose of such studies is to determine whether negative results obtained in germ-cell mutagenicity tests are due to failure of the compound to reach the germ cells. The following methods have been used for such purposes.

1. Unscheduled DNA Synthesis

Unscheduled DNA synthesis (UDS) (as opposed to scheduled DNA synthesis that normally occurs in the S phase of the cell cycle) assays have been used in a number of rodent tissues to provide evidence of induced DNA damage. This endpoint is based on the fact that damage induced in DNA may be repaired by a process that requires resynthesis of regions of DNA where damage occurred. If tritiated thymidine is provided to cells undergoing this repair synthesis, then repair can be detected by autoradiographic or liquid scintillation counting techniques. The methods are suitable for *in vitro* studies, *in vivo* studies, or a hybrid approach in which whole animals are treated with the test chemical, and cells are isolated from the test animal and incubated with tritiated thymidine, followed by determinations of tritium incorporation. All three methods have been used in male germ cells, but only the *in vivo* and the *in vivo/in vitro* will be described here.

The *in vivo* approach involves administration of the test agent to the animals, followed immediately by injection of tritiated thymidine directly into the testes. Spermatozoa are then recovered from the vas deferens or the cauda epididymis, and tritium disintegrations are measured by liquid-scintillation counting. The interval between treatment and spermatozoa recovery can be varied to assess UDS in germ cells at various stages of development. The last round of scheduled DNA synthesis takes place in preleptotene spermatocytes, so spermatozoa collected more than about 35 days following treatment will not yield information on UDS because they will have incorporated tritiated thymidine during S phase.

A variation of this method involves the treatment of animals with the test agent followed by isolation of post-S-phase spermatocytes and spermatids. These cells are incubated with tritiated thymidine, thereby permitting UDS and tritium incorporation to occur *in vitro.* Cells are fixed and slides are prepared and dipped in autoradiographic emulsion and allowed to develop for 3–6 weeks. Slides are scored microscopically by first locating appropriate cell types and then counting silver grains over the nuclei.

In another approach to detect the presence of a test agent or a metabolite in the gonads, radiolabeled test chemical is administered to the animals. By autoradiography or scintillation counting, one can then determine if the test compound has reached the gonads. Where specific cell stages can be isolated, it is possible to determine which cells contain the test agent; by further isolations, one can determine if the compound is bound to specific macromolecules such as DNA or protamines. Although not suitable for routine testing of compounds, information from such studies is highly informative in answering questions as to the presence of test agents in the gonads, cell stages affected, and target molecules. The technique is especially useful in attempts to relate the amount of test chemical reaching the gonads or target cells to the level of effects observed in tests for induced heritable damage.

I. Aneuploidy Tests

A particular type of human genetic disorder that is familiar to most people is aneuploidy. The human diploid chromosome number is 46, and any deviation from this is usually lethal, resulting in death of the conceptus soon after fertilization. In the case of chromosome 21, however, an additional chromosome is often compatible with life, and the individual with three copies of chromosome 21 has a total chromosome number of 47. Genetically, this is termed trisomy 21, and the condition is known as Down's syndrome, readily recognizable because of the characteristic facial features. The incidence of Down's syndrome is roughly 1 in 800 among live births, whereas approximately 1 in 300 liveborns carry aneuploidy of some chromosomes.

At this time, our knowledge of the induction of aneuploidy by environmental agents is restricted by both the limited test methods available and the very limited data base that has accumulated on studies employing mammalian germ-cell systems. Emphasis on this facet of germ-cell genetic toxicity has increased recently, and the number of chemicals studied is increasing. Methods for studying aneuploidy involve assessment of chromosome numbers in developing germ cells, mature germ cells, embryos, or live offspring.

Aneuploid cells can arise during either mitotic or meiotic divisions. The condition results from the unequal distribution of chromosomes to daughter cells and gives rise to cells with extra chromosomes (hyperploid) or cells that have too few chromosomes (hypoploid). In the generation of gametes through meiosis, unequal distribution can occur through nondisjunction of homologous pairs of chromosomes during meiosis I (MI) or nondisjunction of chromatids at meiosis II (MII). In the male, MI nondisjunction leads to all four products of meiosis being aneuploid, whereas MII nondisjunction theoretically gives two normal and two aneuploid gametes.

Mechanisms of aneuploidy induction are not well understood, but it is known that events other than simple nondisjunction can lead to aneuploidy (e.g., the failure of homologous chromosomes to pair during MI).

Although the occurrence of aneuploidy in mitotic spermatogonial cells can be determined cytogenetically, spermatocytes and later cell stages give the most relevant data with regard to aneuploid gametes. Assuming the primary spermatocytes to be normal diploid cells, the array of chromosomes at MI can provide an indication of ensuing aneuploidy. At this stage, homologous chromosomes are normally paired to form bivalents. Unpaired chromosomes are termed univalents. The presence of univalents at MI metaphase suggests a potential for missegregation of chromosomes at anaphase I, but the limited data available do not fully support this association.

A related observation involves the formation of chiasmata (points at which homologous chromatids are in contact following pairing of homologous chromosomes in meiosis) between paired homologs. A reduction in the number of chiasmata, or their absence, can result in the premature separation of homologous chromosomes, with the resulting opportunity for missegregation during anaphase I, leading to aneuploid secondary spermatocytes.

A far less equivocal cytogenetic measure of germ-cell aneuploidy can be obtained by determination of chromosome numbers in metaphases of secondary spermatocytes. At this stage, aneuploidy that arose in MI can be determined with confidence if both hypoploid and hyperploid cells are carefully scored. The importance of hyperploid cells must be stressed because of the possibility that hypoploid cells could result from the artifactual loss of chromosomes during preparation of slides. The scoring of MII cells reflects aneuploidy arising in MI but gives no indication of aneuploidy that could arise during subsequent stages, that is, anaphase II, of the second meiotic division.

To assess the ultimate frequency of aneuploid sperm, it is necessary to determine chromosome numbers in the mature sperm, the male pronucleus in the fertilized egg, or in the first divisions of the zygote.

Two methods have been proposed for detecting aneuploidy in mature sperm. Both are based on unique staining characteristics of heterochromatic regions of sex chromosomes. In the vole *Microtus oeconomus,* both sex chromosomes have an intensely staining region. In Giemsa-stained sperm, be they X or Y, stained regions (assumed to be the sex chromosomes) can often be observed. The presence of zero or two such regions would thus indicate aneuploidy for the sex chromosomes. A similar technique has been used in human sperm, where a region of the Y chromosome fluoresces brightly when stained with the fluorochrome quinacrine dihydrochloride. The presence of two fluorescent bodies suggests an aneuploid condition with two Y chromosomes. This method has been referred to as the YFF test (meaning two fluorescent Y chromosomes). Unfortunately, neither of these two methods has been adequately confirmed genetically and presently cannot be considered for routine use in detecting aneuploidy.

The analysis of developing or mature sperm can be highly informative but does not address the issue of the transmissibility of the aneuploid condition to the subsequent generation. This issue can be investigated by analyzing zygotes and pre- or postnatal offspring.

The chromosome number of sperm that have successfully fertilized eggs can be determined by cytogenetic analysis. At this stage, the maternal and paternal pronuclei can be distinguished by the different degree of condensation of their chromosomes. This method can employ either *in vitro* fertilization or *in vivo* fertilization followed by surgical recovery of the zygotes. Because monosomies and trisomies are usually lethal at some stage of development, zygote analysis permits the detection of a wider range of aneuploidies than analysis at later developmental stages. It is, however, feasible to detect aneuploidy by cytogenetic analysis on somatic tissues at any stage of development.

Finally, the induction or occurrence of aneuploidy in male germ cells can be determined by treating and breeding followed by analysis of progeny. There are two basic methods to detect aneuploidy in the F_1. The first is simply to determine chromosome number through metaphase analysis of somatic cells. In the second method, strains of mice have been developed with genetic markers on the X chromosome that result in certain classes of aneuploid offspring being phenotypically distinct.

Despite the impact on human health, there is still a paucity of information on mechanisms of spontaneous and induced aneuploidy, but research in this area of germ-cell genetic toxicity is increasing. New technologies that permit fluorescent staining of centromeres and the use of specific DNA sequences that can be used to fluorochrome-label sperm chromosomes promise to increase our knowledge in this important area of germ-cell toxicology.

In conclusion, a human health risk associated with exposure to germ-cell mutagens is predicted by our current knowledge of genetics, mutagenesis, and genetically based disease. Further, test systems utilizing laboratory rodents are available for detecting all major classes of genetic damage in germ cells. However, considerable work remains to be done in improving the efficiency and relevance of animal test systems, increasing the accuracy of risk estimates, and better understanding the incidence and origins of genetically based human diseases.

Bibliography

General

Committee on Chemical Environmental Mutagens (1983). "Identifying and Estimating the Genetic Impact of Chemical Mutagens." National Academy Press, Washington, D.C.

Dellarco, V. L., Voytek, P. E., and Hollaender, A., eds. (1985). "Aneuploidy: Etiology and Mechanisms." Plenum, New York.

de Serres, F. J. and Sheridan, W., eds. (1983). "Utilization of Mammalian Specific Locus Studies in Hazard Evaluation and Estimation of Genetic Risk." Plenum, New York.

Drake, J. W. (1975). Environmental mutagenic hazards. Report of Committee 17, Environmental Mutagen Society. *Science* **187,** 503–514.

Russell, L. B., and Shelby, M. D. (1985). Tests for heritable damage and for evidence of gonadal exposure in mammals. *Mutat. Res.* **154,** 68–84.

Carlson, E. A. (1981). "Genes, Radiation, and Society: The Life and Work of H. J. Muller." Cornell Univ. Press, Ithaca, New York.

Test Methods

Dominant Lethal

Anderson D. (1984). The dominant lethal test in rodents. *In* "Mutagenicity Testing—A Practical Approach" (S. Venitt and J. M. Parry, eds.), pp. 307–335. IRL Press, Oxford.

Green, S., Auletta, A., Fabricant, J., Kapp, R., Manandhar, M., Sheu, C., Springer, J., and Whitfield, B. (1985). Current status of bioassays in genetic toxicology—the dominant lethal assay: A report of the U.S. Environmental Protection Agency Gene-Tox Program. *Mutat. Res.* **154,** 49–67.

Heritable Translocation

Generoso, W. M., Bishop, J. B., Gosslee, G. D., Newell, G. W., Sheu, C. J., and Von Halle, E. (1980). Heritable translocation test in mice. *Mutat. Res.* **76,** 191–215.

Morphological Specific Locus

Jenkins, N. A., Copeland, N. C., Taylor, B. A., and Lee B. K. (1981). Diluted coat colour mutation of DBA/2J mice is associated with the site of integration of an ecotropic MuLV genome. *Nature (London)* **293,** 370–374.

Oakberg, E. F. (1956). Duration of spermatogenesis in the mouse and timing of stages of the cycle of the seminiferous epithelium. *Am. J. Anat.* **99,** 507–516.

Russell, L. B., and Matter, B. E. (1980). Whole-mammal mutagenicity tests: Evaluation of five methods. *Mutat. Res.* **75,** 279–302.

Russell, L. B., Selby, P. B., Von Halle, E., Sheridan, W., and Valcovic, L. (1981). The mouse specific-locus test with agents other than radiation: Interpretation of data and recommendations for future work. *Mutat. Res.* **86,** 329–354.

Biochemical Specific Locus

Johnson, F. M., and Lewis, S. E. (1983). The detection of ENU-induced mutants in mice by electrophoresis and the problem of evaluating the mutation rate increase. *In* "Utilization of Mammalian Specific Locus Studies in Hazard Evaluation and Estimation of Genetic Risk" (F. J. de Serres and W. Sheridan, eds.), pp. 95–108. Plenum, New York.

Recessive Lethal

Roderick, T. H. (1983). Using inversions to detect and study recessive lethals and detrimentals in mice. *In* "Utilization of Mammalian Specific Locus Studies in Hazard Evaluation and Estimation of Genetic Risk" (F. J. de Serres and W. Sheridan, eds.), pp. 135–167. Plenum, New York.

Sheridan, W. (1983). The detection of induced recessive lethal mutations in mice. *In* "Utilization of Mammalian Specific Locus Studies in Hazard Evaluation and Estimation of Genetic Risk" (F. J. de Serres and W. Sheridan, eds.), pp. 125–134. Plenum, New York.

Dominant Mutations

Ehling, U. H. (1983). Cataracts—indicators for dominant mutations in mice and man. *In* "Utilization of Mammalian Specific Locus Studies in Hazard Evaluation and Estimation of Genetic Risk" (F. J. de Serres and W. Sheridan, eds.), pp. 169–190. Plenum, New York.

Selby, P. B. (1983). Application in genetic risk estimation of data on the induction of dominant skeletal mutations in mice. *In* "Utilization of Mammalian Specific Locus Studies in Hazard Evaluation and Estimation of Genetic Risk" (F. J. de Serres and W. Sheridan, eds.), pp. 191–210. Plenum, New York.

Selby, P. B., and Niemann, S. L. (1984). Non-breeding-test methods for dominant skeletal mutations shown by ethylnitrosourea to be easily applicable to offspring examined in specific-locus experiments. *Mutat. Res.* **127,** 93–105.

Cytogenetics

Preston, R. J., Au, W., Bender, M. A., Brewen, J. G., Carrano, A. V., Heddle, J. A., McFee, A. F., Wolff, S., and Wassom, J. S. (1981). Mammalian *in vivo* and *in vitro* cytogenetic assays: A report of the U.S. EPA Gene-Tox Program. *Mutat. Res.* **87,** 143–188.

Adler, I-D. and Brewen, J. G. (1982). Effects of chemicals on chromosome aberration production in male and female germ cells. *In* "Chemical Mutagens: Principles and Methods For Their Detection" (F. J. de Serres and A. Hollaender, eds.), Vol. 7, pp. 1–35. Plenum, New York.

Gonadal Exposure

Sega, B. A., and Owens, J. G. (1978). Ethylation of DNA and protamine by ethyl methanesulfonate in the germ cells of male mice and relevancy of these molecular targets to the induction of dominant lethals. *Mutat. Res.* **52,** 87–106.

Sega, G. A. (1974). Unscheduled DNA synthesis in the germ cells of male mice exposed *in vivo* to the chemical mutagen ethyl methansulfonate. *Proc. Natl. Acad. Sci. U.S.A.* **71,** 4955–4959.

Sega, G. A., and Sotomayor, R. E. (1982). Unscheduled DNA synthesis in mammalian germ cells—its potential use in mutagenicity testing. *In* "Chemical Mutagens: Principles and Methods for Their Detection" (F. J. de Serres and A. Hollaender, eds.), Vol. 7, pp. 421–445. Plenum, New York.

Working, P. K., and Butterworth, B. E. (1984). An assay to detect chemically induced DNA repair in rat spermatocytes. *Environ. Mutagen.* **6,** 273–286.

Aneuploidy

Allen, J. W., Liang, J. C., Carrano, A. V., and Preston, R. J. (1986). Review of literature on chemical-induced aneuploidy in mammalian male germ cells. *Mutat. Res.* **167,** 123–137.

Bond, D. J., and Chandley, A. C. (1983). "Aneuploidy. Oxford Monographs on Medical Genetics, No. 11," p. 198. Oxford Univ. Press, London and New York.

Dellarco, V. L., Mavourin, K. H., and Waters, M. D. (1986). An introduction to a series of U.S. Environmental Protection Agency special committee reports on testing approaches for the detection of chemically induced aneuploidy. *Mutat. Res.* **167,** 3–7.

Kapp, R. W., Jr. (1979). Detection of aneuploidy in human sperm. *Environ. Health Perspect.* **31,** 27–31.

Lyon, M. F., Ward, H. C., and Simpson, G. M. (1976). A genetic method for measuring non-disjunction in mice with Robertsonian translocations. *Genetic Res.* **26,** 283–295.

Russell, L. B. (1979). Meiotic nondisjunction in the mouse: Methodology for genetic testing and comparison with other methods. *Environ. Health Perspect.* **31,** 113–122.

Tates, A. D. (1979). *Microtus oeconomus* (Rodential), a useful mammal for studying the induction of sex-chromosome nondisjunction and diploid gametes in male germ cells. *Environ. Health Perspect.* **31,** 151–159.

Chapter 10

Application of *In Vitro* Systems in Male Reproductive Toxicology

TIM J. B. GRAY

The British Industrial Biological Research Association
Carshalton, Surrey SM5 4DS United Kingdom

I. Introduction

The male reproductive system is a most important, and not infrequent, target for chemical toxicity. Assessment of the hazard posed by male reproductive toxicants requires sensitive and reliable procedures for their detection and for establishing underlying mechanisms of toxicity. As in other areas of toxicology, current methodology is based on tests in animals, and these were reviewed in earlier chapters.

In recent years, there has been a major expansion of interest in developing alternative methods in toxicology, such as the use of *in vitro* systems, which may reduce or replace the need for tests in animals (Acosta *et al.*, 1985; Golberg, 1986; Grisham and Smith, 1984; Rowan and Goldberg, 1985). This movement has been stimulated by a variety of factors, including the need to generate toxicity data on an ever-increasing number of chemicals, the rapidly escalating cost of toxicity testing, and mounting public pressure to reduce the use of experimental animals. Interest in the use of *in vitro*

PHYSIOLOGY AND TOXICOLOGY
OF MALE REPRODUCTION

systems in particular has been heightened by advances in cell culture methodology and increasing recognition of the potential of such systems for investigating mechanisms of toxic action. Thus, the facility to study individual cell populations from heterogeneous target organs such as the testis provides a potentially powerful tool for probing mechanisms of organ- and cell-specific toxicity.

The value of working with isolated cells and tissues has long been recognized by physiologists investigating male reproductive function. Such an approach permits the study of individual cell populations with a degree of experimental control impossible to achieve *in vivo* because of the complexity of the reproductive system and its regulatory control. For the testis in particular, an extensive range of *in vitro* systems has been developed, and their use has contributed greatly to our understanding of normal testicular function. The aim of this chapter is to consider the application of such systems to study the disruption of male reproductive function by chemical toxicants. Since this use of *in vitro* systems is still in its infancy, the discussion will be conjectural in places and will be restricted largely to the testis, where most experience has been gained. *In vitro* methodology embraces systems ranging from subcellular fractions to isolated intact organs, but in relation to target-organ toxicity most attention has been focused on the use of primary cell cultures, and that will be the emphasis of this chapter.

II. *In Vitro* Systems from the Male Reproductive Tract

Table I lists some of the many different *in vitro* systems that have been developed and studied by male reproductive physiologists. Primary cultures feature prominently in the table, these having the advantage of retaining *in vitro* many of the differentiated characteristics of their *in vivo* counterparts. However, primary cultures are inherently variable and generally have only a limited lifespan, necessitating frequent preparation of fresh cultures. Cell lines are more uniform in their behavior and can be propagated in the laboratory over prolonged periods of time, but they often have the drawback of bearing only a limited resemblance to the cell type from which they originated. Nevertheless, in the case of Leydig and Sertoli cells, a number of cell lines have been developed that do retain certain differentiated characteristics of these cells, such as response to specific hormones.

Whereas a major advantage of systems consisting largely of a single cell type is their relative simplicity, loss of the interactions with other cell types that influence their function *in vivo* is a potentially serious limitation. The role of such functional interrelationships may be approached *in vitro* by coupling studies with individual cell populations to those in which defined

TABLE I

Some *In Vitro* Systems of Potential Value in Male Reproductive Toxicology

System	References
Sertoli cells	Welsh and Wiebe (1975); Steinberger *et al.* (1975); Kierszenbaum and Tres (1981); Mather and Phillips (1984)
Spermatocytes/spermatids	Meistrich *et al.* (1981); Grootegoed *et al.* (1982); Bucci *et al.* (1986)
Sertoli–germ cell cocultures	Tres and Kierszenbaum (1983); Gray *et al.* (1985)
Leydig cells	Aldred and Cooke (1982); Mather and Phillips (1984); Rommerts *et al.* (1985); Molenaar *et al.* (1983)
Testicular macrophages	Yee and Hutson (1985a)
Peritubular myoid cells	Mather and Phillips (1984); Skinner and Fritz (1986)
Sertoli and Leydig cell lines	Mather *et al.* (1982)
Various cocultures/uses of conditioned medium	
Sertoli–Leydig	Perrard-Sapori *et al.* (1986)
Sertoli–peritubular	Skinner and Fritz (1986)
Sertoli–pituitary	Janecki *et al.* (1984)
Leydig–macrophage	Yee and Hutson (1985b)
Seminiferous tubule segments (mixed or defined stages of spermatogenic cycle)	Hagenäs *et al.* (1978); Toppari and Parvinen (1985)
Isolated perfused testis	Lee and Nagayama (1980)
Epididymal epithelial cells	Byers *et al.* (1986)
Seminal vesicle epithelial cells	Kierszenbaum *et al.* (1983)
Prostatic epithelial cells	O'Connor and Sinha (1985)
Pituitary cell/organ cultures	Janecki *et al.* (1984)
In vitro fertilization	Yanagimachi *et al.* (1976); Moore (1986)

populations are cultured together. In addition to cocultures, the role of soluble secreted factors in cell–cell interactions can be assessed by examining the influence of culture medium conditioned by cells of one type on the function of another. At an increased level of complexity, segments of whole seminiferous tubules can be maintained in culture. In these, the intact seminiferous epithelium can retain its integrity, and even continue to differentiate, for up to several days. A refinement of this approach allows portions of tubules at defined stages of the spermatogenic cycle to be isolated: this may be particularly valuable for studying those toxicants whose effects are apparently exerted only on certain specific stages of spermatogenesis. It is also possible to maintain perfused testes in isolation for several hours, and these have been used, for example, to study xenobiotic metabolism in the testis.

However, the usefulness of such preparations in toxicity studies is limited by their technical complexity and short lifespan *in vitro.*

As is evident from Table I, cell populations from the epididymis, seminal vesicle, and prostate have also been studied *in vitro,* although much less extensively than those from the testis. Pituitary cell cultures and hemipituitaries in organ culture have been used to study gonadotrophic hormone release. Combining these with the use of conditioned medium from Leydig or Sertoli cell cultures enables certain facets of the pituitary–testicular axis to be modeled *in vitro.* Finally, the techniques of *in vitro* fertilization may prove valuable for investigating chemical effects on sperm function and relationships between damage to reproductive tissues and fertility.

The examples mentioned illustrate the diversity of available *in vitro* systems and the range of their complexity from isolated cells to intact organs. In some cases, tissues from a variety of different species, including humans, have been examined. Using such systems to generate relevant toxicological data requires careful attention to many factors, including the procedures used for cell isolation and maintenance and the choice of suitable test systems, dose levels, and criteria of toxicity *in vitro.* These points will be considered in subsequent sections.

III. Methodology for Cell Isolation

The functional properties of cell and tissue preparations can be altered substantially by the procedures involved in their isolation and maintenance *in vitro.* The nature and extent of such changes may have a determinant influence on the outcome of toxicity studies, especially on their reproducibility within and between laboratories. Consequently, careful attention to methodology is most important in developing and applying *in vitro* systems in toxicology.

Many different procedures have been described for preparing cell and tissue cultures from the male reproductive tract. This is true both with respect to the range of different systems listed in Table I and the number of different methods for preparing any particular system. A review of these procedures would be inappropriate here, and the reader is referred to the references in Table I for details of particular methods and to several reviews for additional discussion of underlying principles (Meistrich, 1977; Mather and Phillips, 1984; Rommerts *et al.*, 1985).

Isolating cells from heterogeneous tissues usually involves dissociating the tissue into its constituent cell types, followed by separation and purification of the cells of interest. Dissociation is normally achieved by a combination of mechanical and enzymic (e.g., trypsin, collagenase) treatments, and separation by exploiting differences in cell size and density or in the ability to attach to a suitable culture surface. Almost always, the final cell preparation contains a proportion of other cells in addition to the desired one(s); moreover, most preparative procedures result in some degree of cell damage. Assessing the purity of a cell preparation and the extent of cell damage is most important, since both factors can greatly influence the results of studies *in vitro* (e.g., Molenaar *et al.*, 1983). These and many other considerations relevant to the preparation of primary cell systems from the testis are discussed cogently by Mather and Phillips (1984) and Rommerts *et al.* (1985). The following discussion is restricted to aspects of the isolation of Leydig, Sertoli, and germ cells that are important for the effective use of such systems in toxicological research.

A. Leydig Cells

Interstitial tissue can be freed from the seminiferous tubules by mild enzymic digestion of decapsulated testes (Fig. 1), yielding a cell suspension that contains Leydig cells, but also vascular and connective-tissue elements, testicular macrophages, peritubular myoid cells, and germ cells. The proportion of functionally intact Leydig cells in such a preparation may be no more than 10–20%, making it unsuitable for detailed studies of Leydig cell

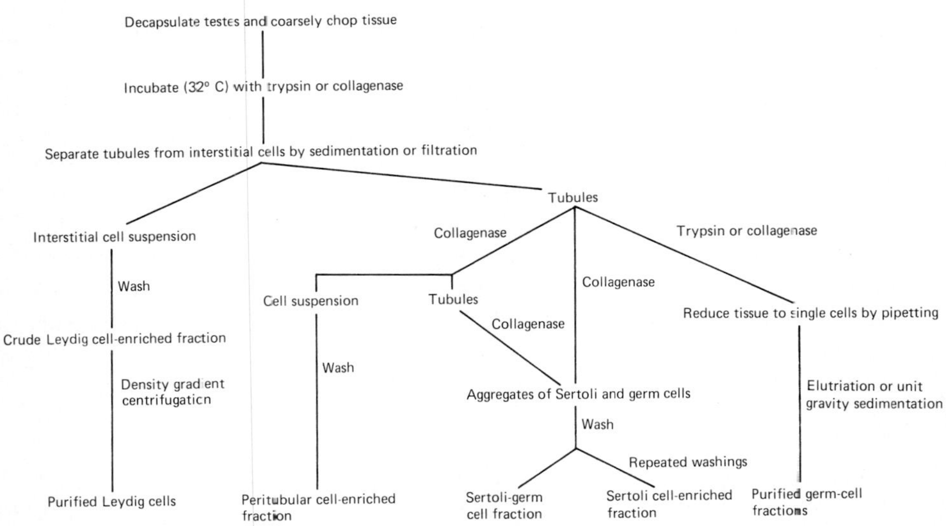

Fig. 1. Procedures for the separation of various cell types from the testis. A combination of mechanical, enzymatic, and sedimentation techniques is used to enrich cell fractions for cell culture.

toxicity. By subjecting the crude cell suspension to density-gradient centrifugation, Leydig cells can be recovered at purities up to 90% or more.

Establishing the percentage of Leydig cells in a preparation requires a means of specifically identifying these cells *in vitro.* The histochemical demonstration of 3β-hydroxysteroid dehydrogenase is most commonly used for this purpose; nonspecific esterase is another suitable marker, though not for Leydig cells from immature rats (Rommerts *et al.,* 1985).

Leydig cells are rather susceptible to damage during isolation. A viable cell count will indicate the proportion of dead cells in the preparation, but apparently viable cells may show varying degrees of functional impairment manifest as a loss of steroidogenic response to luteinizing hormone (LH) stimulation. Measurements of testosterone production in the presence and absence of LH can thus be used as a functional assessment of the "quality" of Leydig cell preparations. These rates vary widely between different laboratories. For adult rat Leydig cells, Rommerts *et al.* (1985) point out that rates of LH-stimulated testosterone production ranging from 3 to 100 ng/10^6 cells per hour have been reported, with the stimulation produced by LH ranging from 4- to 12-fold. This underlines the need for greater standardization of methodology and for the careful characterization of Leydig cell preparations used for toxicity studies. The isolation of two distinct Leydig cell fractions on density gradients has been reported, favoring a view that the rodent testis contains functionally different populations of Leydig cells. However, recent evidence suggests that the lighter of the fractions from density gradients represents damaged cells rather than a distinct subpopulation (Laws *et al.,* 1985).

As well as the actual methodology employed for cell isolation, the yield of functionally intact Leydig cells depends on the age and species of animal from which the testes are taken. In general, higher yields are obtained from testes depleted of germ cells as in immature animals. Immature pig testes are a particularly good source since they contain up to 50% Leydig cells. Testes from animals subject to hypophysectomy or irradiation *in utero* are also suitable, as are Leydig cell tumors, although in these cases the degree to which the cells are representative of cells from mature healthy animals can be questioned.

B. Sertoli Cells

Sertoli cells have been studied in culture more extensively than any other testicular cell type, and many methods for their isolation have been described (see Table I). In principle these all involve sequential dissociation of testes to free the seminiferous tubules of interstitial tissue and peritubular cells and then to break down the tubules to release the Sertoli and germ cells

(Fig. 1). With properly controlled digestion of the tubules, the Sertoli cells remain as aggregates due to the tight junctions between them, and can be separated from the germ cells by repetitive washing and sedimentation (Mather and Phillips, 1984) or by filtration (Steinberger *et al.,* 1975). The washed aggregates will attach to a suitable culture surface and spread to form a monolayer, which can be maintained for up to several weeks.

As inferred above, Sertoli cell cultures are usually contaminated with germ cells and a small proportion of peritubular cells. These contaminants can influence Sertoli cell function (Galdieri *et al.*, 1984a; Skinner and Fritz, 1986), so their numbers should be established. Germ cells and Sertoli cells can readily be distinguished by light microscopy in view of their characteristic and quite different morphology. It is less easy to distinguish peritubular cells from Sertoli cells in this way. However, Chapin *et al.* (1987) have reported that the histochemical demonstration of alkaline phosphatase can be used to identify peritubular cells in culture.

Various strategies have been adopted to minimize germ-cell contamination of Sertoli cell cultures. The most common ploy is to start with testes from very young animals, such as 10- to 18-day-old rats. However, Sertoli cell function changes substantially during sexual maturation (see Chapter 2), and these changes are reflected in the behavior of Sertoli cells *in vitro* (Jutte *et al.,* 1983; Rich *et al.,* 1983). Consequently, the results of *in vitro* studies with chemicals whose testicular toxicity is age-dependent may be biased by the use of very young animals for cell isolation. Germ-cell contamination can also be minimized by using testes depleted of germ cells by hypophysectomy or prenatal irradiation, but Sertoli cells from such testes cannot be considered normal. Alternatively, germ cells can be removed from Sertoli cell cultures, once established, by a brief treatment with hypotonic Tris buffer (Galdieri *et al.,* 1981; Wagle *et al.,* 1986). This procedure has no appreciable effect on Sertoli cell function and is particularly useful for studies with Sertoli cells from older animals, when initial germ-cell contamination will be heavy.

No systematic comparisons have been made of the influence of different isolation procedures on Sertoli cell function *in vitro.* As in the case of Leydig cells, however, lack of standardized methodology may hinder the comparison of toxicity data from different laboratories.

C. Germ Cells

Individual germ-cell populations can be separated from dissociated seminiferous tubules by exploiting differences in cell size and to a lesser extent density (Romrell *et al.,* 1976; Meistrich, 1977; Meistrich *et al.,* 1981; Bucci *et al.,* 1986). The methods most commonly used to achieve this are velocity sedimentation at unit gravity (the Staput procedure) and centrifugal elutriation. Meistrich (1977) has discussed the application and relative merits of these techniques. In both cases, the cell yield and purity obtained depend on the age of animal used and the way in which the initial cell suspension is prepared (Romrell *et al.,* 1976; Bucci *et al.,* 1986). Most success has been achieved in purifying pachytene spermatocytes and early spermatids from mature mouse and rat testes, both procedures giving purities of 80–90%. With further purification by equilibrium density centrifugation, Meistrich *et al.* (1981) recovered fractions containing pachytene spermatocytes and early spermatids as pure as 98 and 93%, respectively, with yields in excess of 10^7 cells. Purification of germ cells at earlier stages of development has proven more difficult, but using immature rats, Bucci *et al.* (1986) have reported the isolation of fractions enriched in type A and type B spermatogonia, zygotene/early pachytene spermatocytes, and midpachytene spermatocytes. The purity of germ-cell fractions is usually assessed by cytological examination of cell smears, but additional confirmation of cell identity can be obtained by determination of cell volume and DNA content.

Isolated germ cells do not attach to culture substrates and remain viable *in vitro* for only a few days at best. Their survival requires addition of lactate and pyruvate to the culture medium, since these cells cannot utilize glucose efficiently (Jutte *et al.,* 1983). Other factors promoting germ-cell survival *in vitro* are actively being sought. Sertoli cell products are likely to have a major role, since germ cells do survive, and may even continue to develop, when cocultured with Sertoli cells (Tres and Kierszenbaum, 1983). Cocultures of Sertoli and germ cells may have potential as an *in vitro* screen for testicular toxicants, and they can be prepared by modifying Sertoli cell isolation procedures so as to leave large numbers of germ cells in the final preparation (Gray *et al.,* 1985) or by adding suspensions of mixed germ cells (Tres and Kierszenbaum, 1983), or purified spermatocytes or spermatids (D'Agostino *et al.* 1984) to Sertoli cell monolayers.

D. Influence of Culture Conditions on Primary Cells

Following isolation, the behavior of cells is further influenced by the conditions used for their maintenance *in vitro.* Important variables include the composition of the culture medium (especially the presence or absence of serum and the addition of hormones and growth factors), the nature of the substrate (e.g., polystyrene dishes, dishes coated with collagen, laminin, fibronectin, extracellular matrix), the plating density, and the temperature of incubation. Incubation time is also an important consideration, since even under "optimal" conditions, primary cells gradually lose differentiated functions and may acquire characteristics unrepresentative of their counterparts *in vivo*. The influence of culture conditions on testicular somatic cell function has been cogently discussed by Mather and Phillips (1984). As yet, however, very little is known about interactions between the variables mentioned above and toxicant action *in vitro*. Generating such information will be a most important task, especially since culture conditions often vary widely between different laboratories.

IV. Application of *In Vitro* Systems in Toxicity Studies

Recent years have seen a very rapid growth of interest in applying *in vitro* methodology to the evaluation of chemical toxicity. Impressive progress has been made in certain areas—for example, in developing and validating *in vitro* tests for genotoxicity and in the development of whole-embryo and other culture systems for identifying teratogens (Golberg, 1986; Rowan and Goldberg, 1985). However, with the possible exceptions of hepatotoxicity and skin and eye irritancy testing, *in vitro* models of target organ toxicity are still in an early stage of development. In the case of the male reproductive system, the wide range of cell and tissue preparations that have been isolated, and that could be developed for toxicity studies, has already been emphasized. Toxicologists have to decide which system(s) to use and how to evaluate toxicity *in vitro*. There are many other problems of experimental design and interpretation to be faced, but a particular one is the selection of appropriate dose levels. These issues are discussed in this section.

A. *Choice of an* In Vitro *System*

The choice of an appropriate *in vitro* system will depend largely on the use to be made of it. In broad terms, this is likely to be either in screening chemicals for potential toxicity or in investigating the mechanisms by which recognized toxicants exert their effects. For mechanistic studies, the choice of system will normally be based on the results of *in vivo* work characterizing the site and nature of tissue damage. Where little is known of the likely mode of action of a chemical, it is desirable to use *in vitro* models that resemble as closely as possible the behavior of the corresponding cells *in vivo*. This generally means using primary cultures. However, where investigations are focused on specific biochemical processes, a cell line possessing the required characteristics could perhaps be used and might help reduce problems due to the variability of primary cultures.

The choice of *in vitro* systems to screen for potential reproductive toxicants is much more problematical, in view of the multiplicity of target sites at which chemicals could act *in vivo* (see Section V). This suggests that systems based on single cell types are unlikely to have broad utility. Thus, primary Leydig cells might provide an effective means of detecting potential Leydig cell toxicants, but there is no reason to suspect that they would be any more useful for detecting spermatocyte toxicants, for example, than any other unrelated cell system. A range of systems could be used as a "test battery," as in genotoxicity testing, but in the present context the value and certainly the practicality of such an approach are questionable. Arguably, the more complex systems such as isolated seminiferous tubule segments or Sertoli–germ cell cocultures may have most potential for toxicity screening. These contain a number of different cell populations, and some of the important interactions between them are retained. At present, however, there is essentially no experience on which to base the choice of *in vitro* systems for screening. Understanding the mechanisms by which chemicals exert their toxicity provides the best basis for this, but in the absence of such understanding, it is likely that working with systems that resemble most closely their *in vivo* counterparts will make it easier to interpret observations *in vitro* in relation to toxicity *in vivo.* Test systems may also be based on empirical correlations between particular effects produced *in vitro* and toxicity *in vivo.* A wide variety of cell systems could be used for this type of approach. However, because the basis of such correlations is, by definition, not understood, the reliability of tests of this sort can only be assessed by extensive validation studies using a diverse range of known toxicants and nontoxicants.

Among more general considerations, the age and species of animal from which tissue is taken are important variables when working with primary cell systems. The anatomy of the testis varies among different species, and its cellular composition changes profoundly between birth and sexual maturity. These factors affect cell behavior, the yield and purity with which individual cell populations can be isolated, and the cellular composition of systems such as Sertoli–germ cell cocultures. Furthermore, pronounced age and species differences in toxicity have been reported for a number of chemicals. For example, in mature rats ethane dimethyl sulfonate is highly toxic to Leydig cells, but it has no such effect in immature rats (Molenaar *et al.*, 1985). Phthalate esters cause severe testicular injury in rats but have little effect in hamsters (Gray *et al.*, 1982). Since toxicology is ultimately concerned with human hazard, the use of *in vitro* systems derived from human tissue is an important possibility. However, it will first be necessary to establish how far known species and age differences in laboratory animal responses to toxicants can be reproduced in corresponding *in vitro* systems.

B. Endpoints of Toxicity In Vitro

The successful application of *in vitro* systems for either screening or mechanistic studies depends on the availability of appropriate criteria by which to evaluate toxicity. A wide range of essentially nonspecific endpoints has been developed in studying the cytotoxicity of chemicals to cells in culture, including the use of vital stains to determine cell viability, measuring the leakage of cytoplasmic enzymes as an index of membrane damage, and the assessment of morphological changes. Such end points are applicable to most *in vitro* systems but tend to reflect only late events in cell injury. For the investigation of cell- and organ-specific toxicity with differentiated cell systems, end points based on specific functional characteristics should prove more sensitive and allow better interpretation of the relationships between *in vitro* and *in vivo* toxicity. Some of the many possible end points that could be developed are listed in Table II. For convenience their application will be considered in relation to specific *in vitro* systems that have been studied in the context of male reproductive toxicology.

1. Leydig Cells

The steroidogenic function of the Leydig cells and the dependence of spermatogenesis on testosterone were discussed earlier (Chapters 2 and 4). Either as freshly isolated cell suspensions or primary cultures, Leydig cells

TABLE II

***In Vitro* Endpoints for Evaluating Male Reproductive Toxicity**

Endpoint	Specific measurements
Secretory products	Leydig cells: testosterone, prostaglandins Sertoli cells: proteins (e.g., APB, transferrin), lactate, pyruvate Pituitary: LH, FSH
Hormone responsiveness	Regulation of secretory products, cAMP response to LH or FSH, receptor binding, protein phosphorylation
Nutrient uptake	Glucose, amino acids
Macromolecule synthesis	DNA, RNA, protein, qualitative changes in protein synthesis (e.g., 2-D PAGE)
Energy metabolism	Carbohydrate and fatty acid utilization
Marker enzymes	Cell specific, subcellular markers
Cell interactions	Germ-cell detachment/attachment, responses to conditioned media
Morphology	Light and electron microscopy, histochemistry, cytochemistry, immunohisto/cytochemisty, fluorescent probes
DNA damage	DNA repair, cytogenetic analysis
Sperm function	Motility, velocity, *in vitro* fertilization, penetration of zone-free hamster ova

continue to secrete testosterone for a limited period *in vitro* and remain responsive to the stimulation of this function by LH or human chorionic gonadotropin (hCG). Measurements of testosterone production, basal and hormone-stimulated, can therefore be used as an index of Leydig cell function *in vitro.* For example, ethanol, which depresses plasma testosterone *in vivo,* produced a dose-related inhibition of testosterone production by suspensions of hCG-stimulated rat Leydig cells, although it did not affect basal production rates (Cicero *et al.,* 1980). Acetaldehyde, the primary metabolite of ethanol, behaved similarly but was around 1000 times more potent an inhibitor than ethanol. Ethane dimethyl sulfonate, a specific Leydig cell toxicant *in vivo* (Molenaar *et al.,* 1985), and certain nonsteroidal antiinflammatory drugs (Dix *et al.,* 1984) are further examples of chemicals that inhibit hormone-stimulated, but not basal testosterone production. In contrast, Burstein *et al.* (1979) showed that Δ^1-tetrahydrocannabinol (THC) inhibited both basal and LH-stimulated testosterone production by mouse Leydig cell suspensions. This was observed at micromolar concentrations considered realistic in relation to the human use of cannabis, which is associated with a decrease in plasma testosterone.

For chemicals that affect hormone-stimulated steroidogenesis, measurements of receptor number and occupancy may be informative, while monitoring changes in cAMP levels following hormone stimulation could illuminate effects on the coupling of receptor occupancy to activation of adenyl cyclase. Measurement of other Leydig cell products, such as prostaglandins and β-endorphin, may provide probes (see Chapter 4) for other aspects of Leydig cell function, but no experience has yet been gained with the application of end points such as these.

Where a chemical is shown to affect testosterone production, isolated Leydig cells can be used to localize the site of its action on the steroidogenic pathway. Laskey *et al.* (1986) proposed a model in which chemical effects are assigned at the level of the plasma membrane receptor or as postreceptor, premitochondrial or postmitochondrial, by measuring effects on LH-, cyclic AMP-, 21-hydroxycholesterol-, and pregnenolene-supported testosterone production, respectively. This type of approach can be extended to dissect the pathway more finely. For example, Burstein *et al.* (1979) localized the inhibitory action of THC to the cholesterol esterase that liberates cholesterol from its cytoplasmic storage pools.

Leydig-cell function *in vivo* is regulated both by LH and local factors, probably originating from the seminiferous tubules (see Chapter 4). Consequently, when evidence of Leydig cell dysfunction is encountered in toxicity studies *in vivo,* it can be difficult to establish whether this is the result of a direct chemical action on the Leydig cells or a secondary consequence of toxicity to the seminiferous tubules or the hypothalamic–pituitary axis.

Using isolated Leydig cells and appropriate cocultures or conditioned medium, it may be possible to distinguish between these various possibilities. However, there are potential pitfalls in this type of approach. For example, Benhamed *et al.* (1982) found that estrogens inhibited testosterone production by crude, but not by purified, preparations of Leydig cells. Similarly, seminiferous tubule-conditioned medium inhibited testosterone production by crude interstitial cell preparations but had a stimulatory effect on Percoll-purified Leydig cells (Syed *et al.,* 1985). These examples reemphasize the points made earlier, that Leydig cell function *in vitro* can be markedly influenced by the preparative methodology employed and that careful characterization of cell preparations is very important.

2. Sertoli Cells

The pivotal role played by Sertoli cells in regulating the development and functions of the testis (see Chapter 2) makes them a potentially important target for chemical toxicity. Several groups have examined toxicant effects on Sertoli cells in primary culture using measurements of Sertoli cell secretory products as indices of toxicity.

In a glucose-containing medium, Sertoli cells secrete lactate and pyruvate in a ratio of approximately 4 : 1, a process that is stimulated by follicle-stimulating hormone (FSH) (Jutte *et al.,* 1983). A number of toxicants have been shown to modulate this function. For example, Beattie *et al.* (1984a) reported that 2-methoxyacetic acid (MAA), an active metabolite of the potent testicular toxicant 2-methoxyethanol, inhibited lactate production by cultured rat Sertoli cells, and they suggested that this might be relevant to the production of spermatocyte injury by MAA and 2-methoxyethanol *in vivo.* In contrast to MAA, however, certain phthalate esters, which are known to be selectively toxic to Sertoli cells, caused a stimulation of lactate production when their active metabolites were added to Sertoli cell cultures. This effect was produced only by those esters causing testicular injury *in vivo* and was evident over a concentration range *in vitro* encompassing the tissue levels likely to be achieved *in vivo* (Moss *et al.,* 1986). Although levels of lactate in the medium were increased, there was no parallel increase in pyruvate, resulting in a marked shift in the lactate/pyruvate ratio. Gossypol, 2,5-hexanedione, 1,2-dibromo-3-chloropropane, cadmium, lead, and dinitrobenzene are other testicular toxicants that have been reported to increase lactate production by Sertoli cell cultures. Measurements of lactate and pyruvate levels in the culture medium thus provide simple, quantitative means of demonstrating effects *in vitro,* but the relationship of these changes to the critical biochemical lesions leading to toxicity *in vivo* is unknown. At present, therefore, such measurements must be regarded as empirical markers, whose wider application will depend on a better understanding of the mechanisms underlying their modulation or on validation studies involving a diverse range of toxicants.

In addition to low-molecular-weight substances, Sertoli cells secrete an extensive range of proteins into the culture medium. The ability of Sertoli cells to survive well in serum-free (i.e., protein-free) culture medium makes the analysis of secretory proteins and their application as potential end points of Sertoli cell toxicity a feasible proposition. Androgen-binding protein (ABP) is probably the best characterized of these proteins, and its use as a functional marker enables observations *in vitro* to be correlated with corresponding measurements *in vivo*. Spitz *et al.* (1985) studied the effects of two indazole carboxylic acid derivatives, lonidamine and tolnidamine, on ABP production by Sertoli cells *in vivo* and *in vitro*. Rats given a single dose of lonidamine had decreased levels of ABP in the testis and epididymis but increased levels in the blood. By treating Sertoli cells in culture, it was shown that lonidamine had no direct effect on ABP production by these cells, leading to the conclusion that *in vivo* it affects primarily the kinetics of ABP secretion from the intact tubule. Zhuang *et al.* (1983) compared the effects of gossypol on primary cultures of Sertoli and Leydig cells. Using ABP and testosterone production, respectively, as functional markers, they reported that Sertoli cells were more sensitive than Leydig cells to the toxicity of gossypol. Studies with Sertoli and Leydig cell lines showed a similar differential sensitivity to gossypol.

Chemical effects on the synthesis of specific proteins, both intracellular and secretory, can be evaluated by the technique of two-dimensional polyacrylamide gel electrophoresis (2-D PAGE). By incubating control and treated cells with appropriate radiolabeled precursors, changes in the pattern of newly synthesized proteins and in protein phosphorylation patterns can be visualized. Very subtle changes can be detected by these techniques, but their specialized nature and potential problems of evaluating the significance of changes observed are likely to restrict their use to specific, mechanistic studies. Nevertheless, in a different but related context, it has been reported that potential teratogens can be identified *in vitro* by their ability to induce characteristic changes in the pattern of protein synthesis by amniotic fluid cells in culture, as judged by 2-D PAGE (Bournias-Vardiabasis, 1985).

The uptake of nutrients and hormones is a potential target for toxic effects on Sertoli cells *in vivo*. Glucose uptake and its regulation by FSH has been studied in Sertoli cell cultures (Hall and Mita, 1984), and the methodology should be applicable to investigations of chemical effects on this, and related processes.

As indicated Chapters 2 and 4, Sertoli cells produce a number of factors that regulate the function of other cell types in both the testis and the pituitary. Chemical effects on these Sertoli-cell functions could be investigated by culturing other testicular cell types or pituitary cells in medium or cytosol preparations from Sertoli cell cultures. This use of *in vitro* systems could provide not only a sensitive means of detecting effects on Sertoli cell function but also a way of distinguishing primary from secondary events in the development of testicular injury. Ample precedent exists in the use of this approach by physiologists elucidating the regulatory interplay between Sertoli cells and Leydig cells (Chapter 4), Sertoli cells and the pituitary (Janecki *et al.*, 1984), and in characterizing the so-called Sertoli cell growth factor (Feig *et al.*, 1980).

Many Sertoli cell functions *in vitro* are responsive to FSH, and valuable additional information may be obtained by investigating chemical effects on both basal and hormone-stimulated function. Where effects on FSH-stimulated cell function are encountered, determinations of receptor number, occupancy, and the cAMP response to FSH binding may provide appropriate end points of toxicity.

The presence of germ cells modulates Sertoli cell function *in vitro*, and responses to the addition of germ cells could provide a further set of end points of toxicity. For example, Sertoli cells form specific attachments with spermatocytes and round spermatids when suspensions of these germ cells are added to Sertoli cell monolayers. Attachment of these germ cells results in stimulation of a variety of Sertoli cell functions, including ABP production (Galdieri *et al.*, 1984a). Changes in germ-cell attachment efficiency following Sertoli cell treatment or changes in the ABP response to germ-cell attachment could provide elegant and sensitive indices of Sertoli cell toxicity.

3. Germ Cells

In chemically induced testicular injury it is often difficult to establish whether germ-cell damage is the result of a direct action on these cells or a consequence of primary toxicity at another site, such as the Sertoli or Leydig cells or the pituitary. Studies with isolated germ cells may help to resolve such problems. However, these cells are comparatively difficult to maintain *in vitro* and, apart from indices such as cell viability, there is a dearth of suitable markers for germ-cell toxicity *in vitro*.

One approach that may be useful is the measurement of energy metabolism in isolated germ cells. For example, Oudiz and Zenick (1986) found that ethoxyacetic acid (EAA) affected respiratory control and decreased cellular ATP levels in isolated pachytene spermatocytes. They speculated that these effects may contribute to the spermatocyte toxicity of EAA *in vivo*. Brabec and co-workers have examined the effect of several toxicants on the utilization of [^{14}C]lactate by isolated spermatocytes (e.g., Beattie *et al.,* 1984b; Miller *et al.,* 1986). Dibromochloropropane and lead both caused dose-related decreases in the conversion of [^{14}C]lactate to $^{14}CO_2$, but MAA, a known spermatocyte toxicant, stimulated lactate oxidation under similar conditions. At present, much more experience is needed to assess the wider utility of these sorts of measurements, although they may be of considerable value in investigating the mechanisms by which individual chemicals exert their toxicity.

A number of studies have shown that the pattern of protein synthesis by germ cells changes both qualitatively and quantitatively as these cells develop through spermatogenesis (e.g., Stern *et al.,* 1983). The identification of proteins whose synthesis is specific to particular stages of germ-cell development may ultimately provide a range of elegant functional markers for germ-cell toxicity both *in vitro* and *in vivo.*

Monitoring Sertoli cell responses to the addition of germ cells was mentioned earlier as a possible index of Sertoli cell toxicity (Section III, B). By treating the germ cells instead of the Sertoli cells, the same approaches could be used as indices of germ-cell toxicity. However, the system might be less easy to work with this way round, since the scope for treating germ cells *in vitro* will be restricted by their limited survival time. A possible solution to this problem could be the use of germ cells isolated from animals pretreated *in vivo.*

4. Sertoli–Germ Cell Cocultures

The ability of germ cells to survive well in coculture with Sertoli cells has been mentioned, as has the potential of such mixed cultures to function as a crude *in vitro* model of the germinal epithelium in screening for potential toxicants.

Using cultures comprising around 20% Sertoli cells and 80% germ cells, Gray and Beamand (1984) examined the effects of a range of toxicants, using as an end point the rate at which germ cells detach from the underlying Sertoli cells. Certain phthalate esters, the indazole carboxylic acid derivative AF1312/TS (Gray and Beamand, 1984) and 1,3-dinitrobenzene (Foster *et al.,* 1987) all produced concentration-dependent increases in germ-cell detachment in this system. No such effects were produced by a number of other phthalate esters that do not affect the testis *in vivo,* while for the toxic phthalates, known age and species differences in their toxicity *in vivo* could be reproduced in the corresponding cocultures. The toxicants mentioned all cause testicular injury characterized by the rapid exfoliation of germ cells from the seminiferous epithelium, and the induction of germ-cell detachment in culture may be a relevant *in vitro* correlate of this action. For chemicals producing different patterns of testicular injury, induction of germ-cell detachment *in vitro* may be a less appropriate end point, Cyclohexylamine, ethyl methane sulfonate, and hydroxyurea, for example, only elicited an appreciable increase in detachment at concentrations that were overtly cytotoxic (Gray and Beamand, 1984).

In a heterogeneous culture system, cell-specific toxicity can be identified by light microscopy of fixed and stained cultures. This has the advantage of enabling changes produced *in vitro* to be related directly to the corresponding histological picture *in vivo.* For example, treatment of Sertoli–germ cell cocultures with MAA results in degeneration of pachytene spermatocytes without affecting the spermatocytes at other stages, spermatogonia, or Sertoli cells (Gray *et al.,* 1985) (Fig. 2). This sort of specific morphological change, restricted to the putative target cell type as identified *in vivo,* has also been seen with phthalate esters and cyclohexylamine. As with MAA, these effects occur at concentrations well below those causing nonspecific morphological changes indicative of overt cytotoxicity. Thus, it appears from the limited experience gained so far that morphological examination of treated cocultures may provide a sensitive means of assessing toxicity *in vitro* that can be related to the histological evaluation of testicular injury *in vivo.*

Morphological studies can readily be extended to the ultrastructural level, to assist identification of subcellular targets of toxicity. Of course this applies not only in the case of Sertoli–germ cell cocultures. Whatever the cell system, electron microscopy can also be used effectively to evaluate just how closely toxicant-induced changes mimic those occurring *in vivo,* before pursuing mechanistic studies *in vitro* (Creasy *et al.,* 1986).

A disadvantage of the use of morphological criteria to evaluate toxicity is the difficulty of quantitating the changes observed. Measuring cell-specific marker enzyme activities is one way in which quantitative data may be obtained from the Sertoli-germ cell coculture system. In their studies with glycol ether metabolites, Gray *et al.* (1985) observed decreases in activity of the pachytene spermatocyte marker carnitine acetyltransferase, which correlated well with production of the morphological changes described earlier. In principle, a range of other functional markers could be used, representative of the various cell populations present in this culture system. Such an approach could be appropriate in the context of toxicity testing, but not necessarily as a means of identifying target cell types, in view of the extensive cell–cell interactions in these cocultures. For example, a decrease in ABP production following treatment might be the result of direct Sertoli cell toxicity or due to an effect on the germ cells, causing their detachment from the Sertoli cells, which in turn depresses Sertoli cell function (Galdieri *et al.,* 1984a). Resolving problems of this nature, which may be unimportant when merely aiming to rank compounds in order of potential toxicity, requires the use of purified preparations of the relevant cell types.

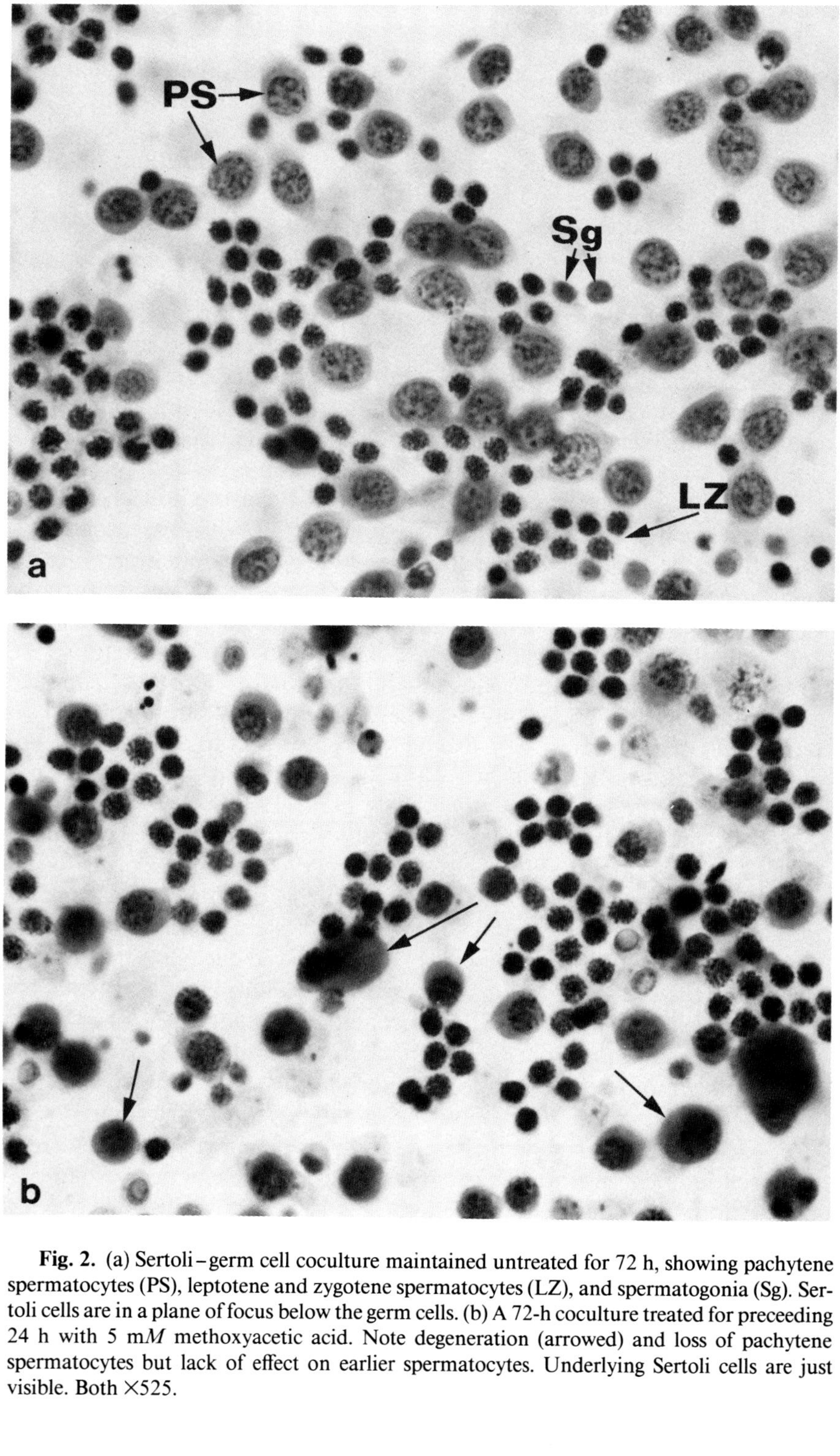

Fig. 2. (a) Sertoli–germ cell coculture maintained untreated for 72 h, showing pachytene spermatocytes (PS), leptotene and zygotene spermatocytes (LZ), and spermatogonia (Sg). Sertoli cells are in a plane of focus below the germ cells. (b) A 72-h coculture treated for preceeding 24 h with 5 m*M* methoxyacetic acid. Note degeneration (arrowed) and loss of pachytene spermatocytes but lack of effect on earlier spermatocytes. Underlying Sertoli cells are just visible. Both ×525.

5. Seminiferous Tubule Cultures

A number of the endpoints discussed earlier may be applicable to cultures of intact segments of seminiferous tubules, such as morphological examination and measurements of Sertoli cell secretory products. These cultures are potentially useful because they combine maximal retention of *in vivo* architecture and function (compared to isolated cell preparations) with the advantages of *in vitro* systems in terms of control of experimental conditions. For in-depth studies of cell-specific toxicity, however, the relative complexity of tubule cultures is likely to limit their utility. The facility to isolate segments of tubules at defined stages of the spermatogenic cycle is potentially a very powerful feature of this culture system. As yet it does not appear to have been applied to the analysis of stage-specific testicular injury.

6. Pituitary Cell and Organ Culture

The use of these systems to measure effects on gonadotrophic hormone production, and to measure pituitary responsiveness to normal stimulatory (gonadotropin-releasing hormone, GnRH) and inhibitory (inhibin) influences, could provide a means of identifying chemicals whose reproductive toxicity is exerted directly at the level of the pituitary. By using cocultures or conditioned medium, effects on interactions between the pituitary and the testis can also be investigated (e.g., Janecki *et al.,* 1984).

7. *In Vitro* Fertilization and Effects on Sperm Function

The technique of *in vitro* fertilization (IVF) has considerable potential as a sensitive indicator of chemical effects on the fertilizing capacity of sperm. The methodology is well established for the mouse and hamster (see review by Moore, 1986), but as yet little experience has been gained in its application in reproductive toxicology. Furthermore, it has been difficult to achieve consistent success with IVF in the rat, the species most commonly used for toxicity studies (Moore, 1986). As well as providing a possible means of assessing effects on fertility, IVF should allow something to be learned of the action of toxicants, since different components of the fertilization process (e.g., acrosome reaction, sperm binding and penetration, sperm – egg fusion) can be observed directly. A further advantage of IVF for the toxicologist is its applicability to human gametes, although substantial ethical problems remain to be resolved.

Other endpoints for assessing toxicant effects on sperm function *in vitro* include measurements of oxygen consumption, marker enzyme activities, sperm motility and velocity, and the penetration of zona-free hamster ova. As in the case of IVF, exposure to the test chemical can be *in vivo* or sperm can be treated directly *in vitro.* The applicability of these techniques to human as well as laboratory animal sperm makes them of particular interest.

For example, the ability of human sperm to penetrate zona-free hamster ova correlates well with fertility *in vivo* (see Moore, 1986), and the use of this approach avoids the ethical problems of IVF. It could be used to evaluate toxicant effects on sperm function both in the laboratory and in monitoring individuals exposed in the workplace or through drug therapy.

8. Other Endpoints of Toxicity

Although the emphasis so far has been on cell-specific endpoints of toxicity *in vitro*, there are many more generally applicable endpoints that may be useful in the context of male reproductive toxicology. Some of these are considered below.

a. Energy Metabolism. Because of the intense metabolic activity associated with spermatogenesis and the relatively low levels of certain glycolytic enzymes in spermatocytes and spermatids, the testis may be particularly vulnerable to chemicals that interfere with energy metabolism. *In vitro* systems lend themselves well to the investigation of such effects. Pathways of carbohydrate and fatty acid utilization can be monitored by measuring oxygen consumption, ATP production, the production of $^{14}CO_2$ from appropriately labeled substrates, and/or the incorporation of label into cellular lipids. For example, lead increased lactate production by cultured Sertoli cells, but decreased the conversion of [^{14}C]glucose to $^{14}CO_2$, suggesting a postglycolytic block in glucose utilization (Batarseh *et al.*, 1986). Pathways can be dissected further, with the use of appropriate radiolabeled metabolic intermediates, to locate more precisely the site of action of chemicals affecting energy metabolism.

b. Macromolecule Synthesis. Measuring rates of DNA, RNA, and protein synthesis based on incorporation of radiolabeled precursors is one of the most widely used procedures in *in vitro* toxicology. Changes in macromolecule synthesis provide relatively simple, quantitative indices of effect, and in the case of proteins, the synthesis of both intracellular and secreted proteins can be measured at the same time. However, measurements of overall macromolecule synthesis tend to be somewhat insensitive. Small changes in total protein synthesis, for example, may conceal marked alterations in the pattern of synthesis of individual proteins. Equally, it is not uncommon to find significant changes in specific cellular functions in the absence of effects on overall macromolecule synthesis. Thus Beattie *et al.* (1984a) reported marked inhibition of Sertoli cell lactate production by MAA at concentrations that had no effect on protein synthesis in these cells. Similarly, lonidamine (0.06 μM) inhibited aromatase activity in Sertoli cells by more than 50% but had no effect on protein or RNA synthesis (Galdieri *et al.*, 1984b).

c. Marker Enzymes. Cell-specific enzymes may be useful to localize toxic effects to a particular cell type when working with heterogeneous cell systems *in vitro.* This was illustrated earlier by the use of carnitine acetyltransferase as a marker for pachytene spermatocytes in Sertoli–germ cell cocultures. In an essentially homogeneous cell system cell specificity may not be important, but enzyme or other markers for different subcellular compartments may help localize effects to a particular site of action. Examples of this sort of approach were discussed earlier in relation to Leydig cell toxicity (Section IV,B,1).

d. Cytochemistry and Immunocytochemistry. These can be powerful tools for the analysis of chemical toxicity at the cellular level. For example, recently introduced fluorescent probes such as rhodamine 123 enable mitochondrial function to be studied in living cells. These dyes are selectively concentrated by normally functioning mitochondria, and this can be visualized by fluorescence microscopy. Loss of specific staining may provide a sensitive index of mitochondrial injury. For example, Tanphaichitr *et al.* (1984) showed that addition of gossypol to rat testicular tumor cells or sperm preloaded with rhodamine 123 led to a very rapid loss of specific mitochondrial staining. Changes in the intracellular distribution and concentration of calcium are thought to have an important role in the regulation of cell function and in mediating cell injury. Such changes can be observed directly *in vitro* using recently introduced calcium-specific fluorescent probes.

Immunocytochemical techniques can be used to visualize a wide variety of cellular constituents and have the advantage of allowing changes observed *in vitro* to be correlated with parallel observations on tissue sections from animals treated *in vivo.* Techniques for demonstrating cytoskeletal elements, for example, have been used to study the effects of FSH on Sertoli cell morphology *in vitro* and may be equally useful for investigating toxicant-induced changes.

e. Cell Viability. Methods of assessing cell viability are widely used in studying the cytotoxicity of chemicals to cells in culture. These methods include the use of vital stains, measuring the leakage of cytoplasmic enzymes into the culture medium as an index of damage to cell membranes, measuring levels of key cellular constituents such as ATP, and light microscopy to identify morphological changes indicative of toxicity (Acosta *et al.,* 1985). Such procedures can identify treatment levels causing lethal cell injury. However, it is changes in specific cell functions produced at levels well below those causing lethal cell injury that are much more likely to be relevant to the production of cell- and organ-specific toxicity.

C. Selection of Dose Levels

For the results of studies *in vitro* to be relevant, and to be interpretable in relation to toxicity and hazard assessment *in vivo,* careful selection of dose levels is very important. There is almost unlimited scope for generating misleading data by the use of chemical concentrations *in vitro* that would never be obtained, or persist, *in vivo.* Indeed, Paracelsus's famous maxim, that all things are toxic depending only on the dose, is particularly apposite in the context of *in vitro* toxicology.

Two practical approaches to this problem can be identified. Ideally, dose levels studied *in vitro* should be based on knowledge of blood and/or tissue levels associated with toxicity *in vivo.* In this way, changes occurring only at much higher levels *in vitro* can be put in their proper context, as being of doubtful relevance. However, the necessary pharmacokinetic data are likely to be available only in the case of chemicals whose toxicity has already been recognized and studied in some detail *in vivo*—in other words, when *in vitro* systems are being used to pursue mechanisms of toxicity.

Where pharmacokinetic data are not available, as in the case of screening untested chemicals for potential reproductive toxicity, dose levels should be chosen so as to allow full dose–response relationships to be established for the end points studied. This provides the best basis for comparing the relative toxic potential of different chemicals *in vitro.* Nevertheless, the toxicological significance of such data can be very difficult to evaluate, a point that is discussed further in Section V.

A number of other, more generally applicable, factors need to be considered when deciding on dose levels. These include the solubility of the test compound and its possible effects on the pH, ionic strength, and osmolarity of the culture medium. In the case of potentially surface-active agents, such as phthalate monoesters, it is important to ensure that the critical micellar concentration is not exceeded if nonspecific effects are to be avoided. Where serum-containing medium is used, protein binding has to be considered, since this may reduce the concentration of free compound available to the cells to well below the nominal concentration.

V. Critical Appraisal of the Usefulness of *In Vitro* Systems

The previous section illustrated some of the many ways in which *in vitro* systems are being, or could be, used to study the effects of male reproductive toxicants. While this approach has undoubted potential, its usefulness in practice depends on the extent to which effects produced *in vitro* can be related to toxicity *in vivo*—in other words, on whether the data generated is relevant and can contribute to hazard assessment. This is a crucial issue, for while it is easy to produce effects by adding chemicals to cells in culture, it may be extremely difficult to establish what relationship, if any, these effects have to the mechanism by which toxicity is produced in the intact animal. These difficulties are at a minimum when the use of *in vitro* systems is restricted to circumstances in which specific, well-defined questions can be asked. For example, when the mechanism of action of a recognized toxicant is being investigated, there will already be a framework of knowledge on which to base experimental design and the interpretation of results. When working with previously untested chemicals, as in toxicity screening, there will be no such framework, and the problems of interpretation may be formidable.

A. In Vitro *Systems and Toxicity Screening*

Ideally, for detecting organ-specific toxicants, *in vitro* systems should be capable of modeling accurately the responses of target tissues *in vivo.* A less satisfactory alternative would be systems whose response could be empirically, but reliably, related to the production of toxicity *in vivo* based on validation studies with chemicals of known toxicity. There are serious obstacles to achieving either of these possibilities in the present context. For example, the susceptibility of the male reproductive tract to toxicants reflects to a large extent the complexity of its structure, function, and regulatory control, and the consequent multiplicity of sites at which toxicants might exert their effects. Much of this complexity and many potential target sites are lost in the preparation of *in vitro* systems. The response of the testis in particular may depend on critical cell–cell interactions within the tissue. These too are modified or disrupted *in vitro.* At present, meiosis and postmeiotic germ cell development cannot be satisfactorily accomplished *in vitro*, precluding studies of chemical effects on these critical steps in spermatogenesis. Even when cell-specific functions are apparently retained *in vitro,* subtle changes in the behavior of cells may alter their response to toxicants. Thus, Sertoli cells in culture express β-adrenergic receptors that are not evident *in vivo,* and the basal rate of testosterone production by isolated Leydig cells is probably much lower than *in vivo.* Cell behavior in culture is also influenced by factors such as plating density and culture-medium composition, especially the presence or absence of added hormones: at present virtually nothing is known of how these factors affect cell sensitivity to toxicants.

As well as these difficulties relating to model development, there are others that apply even if an entirely empirical approach to testing is considered. These include the role of the blood–testis barrier (Chapter 2) in governing access of chemicals to the meiotic and postmeiotic germ cells *in vivo,* the low levels of xenobiotic metabolizing enzymes present in most cell culture systems, and the inability of static *in vitro* systems to model the dynamic processes of absorption, distribution, and excretion that are important determinants of toxicity *in vivo.* In the absence of the blood–testis barrier, germ cells *in vitro* may be freely exposed to chemicals that would be excluded from their environment *in vivo.* In such circumstances, effects observed *in vitro* may have little or no relevance to toxicity *in vivo.* Lack of xenobiotic metabolizing capacity is a feature common to many *in vitro* systems and may result in failure to detect those chemicals whose toxicity is mediated by metabolites rather than the parent compounds themselves. For example, Sertoli–germ cell cocultures showed no response to 2-methoxyethanol or di-(2-ethylhexyl) phthalate, two well-characterized testicular toxicants (Gray and Beamand, 1984; Gray *et al.,* 1985). Only when their respective active metabolites were added to the cultures was toxicity evident. However, Foster *et al.* (1987) were able to demonstrate metabolism of 1,3-dinitrobenzene to a postulated toxic intermediate in Sertoli cell cultures. Nevertheless, if *in vitro* systems are ever to be useful for screening untested chemicals, some means of achieving wider metabolic competence will have to be found, perhaps by using the liver-derived metabolizing systems that have been extensively employed in genetic toxicity testing. The problems of selecting appropriate dose levels for *in vitro* studies were discussed earlier (Section IV,C). These, together with the loss of permeability barriers and clearance mechanisms operating *in vivo,* mean that cells *in vitro* may be exposed to chemicals at concentrations, and for durations, quite unrelated to the conditions obtaining in treated animals.

There are clearly many problems to be overcome before *in vitro* systems can be used in screening as reliable predictors of male reproductive toxicity. However, this should not be dismissed as an unattainable goal in view of the diversity of approaches that could be explored (Section IV,B) and continuing developments in methodology. One almost immediate possibility is that of screening within a particular class of structurally related chemicals that may share a common mode of action although differing quantitatively in their toxicity. The phthalate esters and glycol ethers are examples of such classes. In this situation, the development of an *in vitro* model is based on *in vivo* characterization of the toxicity, and the role of metabolism in the toxicity, of several members of the class. Using such systems to predict the toxicity of other chemicals within the class involves far fewer uncertainties than screening chemicals of diverse structure and unknown toxicity.

B. **In Vitro** *Systems and Mechanistic Studies*

At present the most appropriate applications for *in vitro* systems are as experimental tools for addressing specific questions about the actions of recognized toxicants. Investigations of this nature are not normally undertaken until *in vivo* studies with the chemical in question have provided information on its target cell type(s), on the role of metabolism in its toxicity, and on tissue levels and clearance rates. In these circumstances, many of the limitations discussed in the context of screening no longer apply; indeed, the relative simplicity of *in vitro* systems becomes a positive advantage and their use may provide information that could not readily be obtained *in vivo.*

Distinguishing primary from secondary events in male reproductive toxicology may be very difficult to accomplish *in vivo* because of the extensive interactions between different cell populations. For example, morphological evidence indicates that MAA is preferentially toxic to pachytene spermatocytes, but it is not clear whether this reflects a direct action on those cells or whether it is a consequence of primary damage to the Sertoli cells causing loss of functions critical to normal spermatocyte development. The use of pachytene spermatocytes in isolation, and in coculture with Sertoli cells, provides a means of resolving this issue. However, the success of this approach depends on knowing enough about the effects of MAA on pachytene spermatocytes *in vivo* to ensure that their response to treatment *in vitro* can be properly evaluated.

These principles can be extended to other situations where the influence of cell–cell interactions and regulatory control mechanisms *in vivo* makes it difficult to establish the sequence of events leading to functional impairment or morphologically observable cell injury. Thus, where decreased levels of plasma testosterone are encountered in treated animals, isolated Leydig cells could be used to establish whether the chemical acts directly on these cells. If there is no evidence of toxicity at appropriate dose levels *in vitro,* effects on the paracrine control of Leydig cell function could be probed with the use of conditioned medium from control and treated Sertoli cell cultures.

In addition to their value in identifying primary cellular targets, *in vitro* systems are well suited to the more detailed analysis of toxicant action at the cellular and subcellular level. The ease with which these systems can be controllably manipulated means that the interaction of toxicants with specific metabolic pathways and processes can be dissected in detail and sites of action can be determined. The example of chemicals affecting testosterone production by Leydig cells was mentioned earlier (Section IV, B,1) to illustrate how isolated cells can be used to identify the particular step(s) in the steroidogenic pathway at which a given chemical is acting. It may often be possible to evaluate the relationship of observed biochemical changes to the expression of toxicity *in vitro* by adding specific inhibitors or reaction products to the culture medium. Thus, Beattie *et al.* (1984a) suggested that the spermatocyte toxicity of MAA *in vivo* might be due to deprivation of energy substrates, based on observations that MAA inhibited lactate production by Sertoli cells in culture. However, using Sertoli–germ cell cocultures, Gray *et al.* (1985) showed that supplementing the medium with lactate did not influence the spermatocyte toxicity of MAA.

For toxicants whose metabolic fate has been established, *in vitro* systems can be used to evaluate the relative toxicity of the parent chemical and its individual metabolites. This is particularly important since clarifying the relationship of metabolism to toxicity, and identifying the active toxicant, is a prerequisite for studies on mechanisms of toxicity. In view of the importance of metabolism in toxicity, the distribution of xenobiotic metabolizing enzymes within target tissues may have a significant role in determining the cell specificity of toxicant action. The use of isolated cell populations should enable the distribution of metabolizing capacity within the heterogeneous tissues of the male reproductive tract to be characterized more extensively than is presently the case.

C. Species Differences in Toxicity and the Use of Human Tissue

The toxicity of a number of chemicals is known to vary considerably between different species of laboratory animal. This is usually due to differences in metabolism, pharmacokinetics, and/or target organ sensitivity: data from *in vitro* studies can help distinguish between these possibilities. For example, if the metabolite(s) responsible for toxicity in a sensitive species fails to produce toxicity in cultures from a resistant species, a role for differences in target tissue sensitivity is suggested. Conversely, if it does cause toxicity *in vitro,* the lack of toxicity *in vivo* may reflect a species difference in metabolism or pharmacokinetics such that an insufficient amount of the active metabolite is formed or reaches its site of action.

The existence of species differences in toxicity raises the question of whether effects observed in sensitive animal species would also be produced in humans. The use of *in vitro* systems derived from human tissue provides a potentially very important way of resolving such questions and thereby contributing to human hazard assessment. Although there are obvious difficulties in obtaining normal tissue, particularly from prepubertal individuals, methodology for maintaining preparations of human testis and epididymis *in vitro* has been established and the application of such systems in toxicology has begun to be actively pursued. For the data to be interpretable in terms of human hazard, however, these studies will need to be based on extensive knowledge of the relationship between the *in vitro* and *in vivo* effects of the test chemical in corresponding tissue from sensitive animal species.

D. Further Development of In Vitro *Systems*

Many of the limitations of *in vitro* systems referred to earlier, especially those concerned with deriving accurate models of *in vivo* behavior, may be overcome as progress in reproductive physiology and *in vitro* methodology continues to be made. This will involve both the refinement of existing systems and the exploitation of additional ones such as *in vitro* fertilization.

One area in which significant advances have already been made is in the development of more sophisticated culture substrata, which enhance the retention *in vitro* of differentiated cellular characteristics. For example, when Sertoli cells are cultured on surfaces coated with a reconstituted basement membrane gel, they retain the tall columnar morphology with basally located nucleus that is typical of these cells *in vivo* (Hadley *et al.,* 1985). On conventional tissue-culture plastic, Sertoli cells are flattened and show no nuclear polarity. It remains to be determined how the profound influence of culture substrate on both morphology and cell function affects the response of cells to toxicant action.

Another feature of Sertoli cells cultured on extracellular matrix preparations but not on conventional plastic is the formation of tight junctions between the basal portions of adjacent cells. Such junctions form the principal component of the blood-testis barrier *in vivo.* Using two-compartment culture systems, it can be shown that these junctions occlude passage through the cell layer of substances added to the medium bathing the basal surface of the cells (Byers *et al.,* 1986). This raises the important possibility of modeling the blood-testis barrier *in vitro,* thereby providing a system in which the likelihood of potential toxicants reaching the developing germ cells could be directly assessed. This might be particularly valuable in the case of somatic mutagens, to evaluate the probability that they would also cause genetic damage in the meiotic and postmeiotic germ cells. From a screening standpoint, incorporation of this important permeability barrier into the *in vitro* system should reduce the number of false positives that would be expected in assessing germ-cell toxicity in systems where the chemical has free access to the germ cells.

Another area in which developments should help the toxicologist is in the formulation of improved culture media. As more is learned of the precise conditions necessary for postmeiotic germ-cell development *in vivo,* it may be possible to identify specific factors that will enhance the survival and even the differentiation of isolated germ cells. The ability to culture Sertoli cells under conditions promoting the formation of basal tight junctions raises the prospect of creating an environment *in vitro,* analogous to the adluminal compartment of the seminiferous tubule, in which germ-cell development might take place (see Chapter 2).

The development of perifusion techniques may help overcome some of the limitations posed by the static nature of most *in vitro* systems. Perifusion allows the continual removal and the monitoring of secretory products and metabolites during exposure to chemicals and provides the possibility of reproducing *in vitro* the fluctuating hormonal environment of cells *in vivo.*

Where sufficient is known of compound pharmacokinetics, perifusion could enable conditions of target cell exposure *in vivo* to be closely modeled. Another development could be the use of cells isolated from animals pretreated with the test chemical. This sort of combined *in vivo/in vitro* approach enables metabolic and pharmacokinetic factors influencing toxicity to be taken into account while retaining the advantages of subsequent analysis of toxicity *in vitro.* It has been used with success in genotoxicity testing, but there are potential technical difficulties to be considered. Thus, the effects of treatment may interfere with cell-isolation procedures, and, in any event, it will be virtually impossible to establish cultures from treated and control animals under exactly the same conditions. Furthermore, this approach will not be suitable with systems such as Sertoli cells, which require a significant time for attachment and spreading before they can be subject to analysis *in vitro.*

The foregoing comments are based on the premise that the more "*in vivo*-like" *in vitro* systems can be made to be, the more valuable they are likely to prove. For the toxicologist, however, there must be a limit to the degree of complexity worth building into *in vitro* systems. Beyond that limit it will make more sense on all except perhaps humanitarian grounds to use whole animals.

VI. Conclusions

Current interest in the application of *in vitro* systems in toxicology is likely to persist as the quest continues for methods of assessing toxicity that are quicker, more sensitive, and less dependent on the use of laboratory animals than established procedures. However, there are many problems to be resolved before *in vitro* systems can have a major impact in reducing the burden of animal testing. In the case of the male reproductive tract, difficulties arising from the inability to model metabolic and pharmacokinetic influences on toxicity are compounded by the complexity of the tract, such that most *in vitro* models will contain only a fraction of the many potential sites for toxicant action *in vivo.* The consequent uncertainties involved in the interpretation of results effectively preclude the use of *in vitro* systems for toxicity screening with previously untested chemicals. With developments in cell and molecular biology, some of these difficulties will doubtless be overcome, but the process is likely to be a slow one. Attention will also need to be given to assessing the relationships between tissue damage and the ability to father normal offspring: most of the systems considered in this chapter address only the question of tissue damage. For the immediate future, the best use of *in vitro* systems will be in following up initial studies in animals. In this way, the uncertainties just mentioned can be minimized, and it should be possible to exploit the very considerable potential of *in vitro* systems for elucidating mechanisms of toxicity and generating data on the comparative sensitivity of human and animal reproductive tissues to suspected toxicants. In the longer term, it is progress in our understanding of mechanisms of toxicity that will provide the basis for developing *in vitro* systems that can be used with confidence to predict toxicity *in vivo* and to exploit most effectively the use of human tissues for this purpose.

Acknowledgment

My thanks to Marjorie Turner-Pach for much patient effort in typing this manuscript.

References

Acosta, D., Sorensen, E. M. B., Anuforo, D. C., Mitchell, D. B., Ramos, K., Santone, K. S., and Smith, M. A. (1985). An *in vitro* approach to the study of target organ toxicity of drugs and chemicals. *In Vitro* **21,** 495–504.

Aldred, L. F., and Cooke, B. A. (1982). The deleterious effect of mechanical dissociation of rat testes on the functional activity and purification of Leydig cells using Percoll gradients. *Int. J. Androl.* **5,** 191–195.

Batsarseh, L. I., Welsh, M. J., and Brabec, M. J. (1986). Effects of lead acetate on Sertoli cell lactate production and protein synthesis *in vitro. Cell Biol. Toxicol.* **2,** 283–292.

Beattie, P. J., Welsh, M. J., and Brabec, M. J. (1984a). The effect of 2-methoxyethanol and methoxyacetic acid on Sertoli cell lactate production and protein synthesis *in vitro. Toxicol. Appl. Pharmacol.* **76,** 56–61.

Beattie, P., Miller, G., Welsh, M., and Brabec, M. J. (1984b). Dibromochloropropane and methoxyacetic acid alter lactate metabolism by isolated rat spermatocytes. *Biol. Reprod.* **30,** Suppl. 1, 169, abstr.

Benahmed, M., Bernier, M., Ducharme, J. R., and Saez, J. M. (1982). Steroidogenesis of cultured purified pig Leydig cells: Secretion and effects of estrogens. *Mol. Cell. Endocrinol.* **28,** 705–716.

Bournias-Vardiabasis, N. (1985). Use of amniotic fluid cells for testing teratogens. *In* "*In Vitro* Toxicology" (A. M. Goldberg, ed.), pp. 315–331. Mary Ann Liebert Inc., New York.

Bucci, L. R., Brock, W. A., Johnson, T. S., and Meistrich, M. L. (1986). Isolation and biochemical studies of enriched populations of spermatogonia and early primary spermatocytes from rat testes. *Biol. Reprod.* **34,** 195–206.

Burstein, S., Hunter, S. A., and Shoupe, T. S. (1979). Site of inhibition of Leydig cell testosterone synthesis by Δ^1-tetrahydrocannabinol. *Mol. Pharmacol.* **15,** 633–640.

Byers, S. W., Hadley, M. A., Djakiew, D., and Dym, M. (1986). Growth and characterization of polarized monolayers of epididymal epithelial cells and Sertoli cells in dual environment culture chambers. *J. Androl.* **7,** 59–68.

Chapin, R. E., Phelps, J. L., Miller, B. E., and Gray, T. J. B. (1987). Alkaline phosphatase histochemistry discriminates peritubular cells in rat primary testicular cell cultures. *J. Androl.*, **8,** 155–161.

Cicero, T. J., Bell, R. D., Meyer, E. R., and Badger, T. M. (1980). Ethanol and acetaldehyde directly inhibit testicular steroidogenesis. *J. Pharmacol. Exp. Ther.* **213,** 228–233.

Creasy, D. M., Jones, H. B., Beech, L. M., and Gray, T. J. B. (1986). The effects of two testicular toxins on the ultrastructural morphology of mixed cultures of Sertoli and germ cells: a comparison with *in vivo* effects. *Food Chem. Toxicol.* **24,** 655–656.

D'Agostino, A., Monaco, L., Stefanini, M., and Geremia, R. (1984). Study of the interaction between germ cells and Sertoli cells *in vitro. Exp. Cell Res.* **150,** 430–435.

Dix, C. J., Habberfield, A. D., Sullivan, M. H. F., and Cooke, B. A. (1984). Inhibition of steroid production in Leydig cells by non-steroidal anti-inflammatory and related compounds: evidence for the involvement of lipoxygenase products in steroidogenesis. *Biochem. J.* **219,** 529–537.

Feig, L. A., Bellvé, A. R., Erickson, N. H., and Klagsburn, M. (1980). Sertoli cells contain a mitogenic polypeptide. *Proc. Nat. Acad. Sci. U.S.A.* **79,** 4774–4778.

Foster, P. M. D., Lloyd. S. C., and Prout, M. S. (1987). Toxicity and metabolism of 1,3-dinitrobenzene in rat testicular cell cultures. *Toxic. in Vitro* **1,** 31–37.

Galdieri, M., Ziparo, E., Palombi, F., Russo, M. A., and Stefanini, M. (1981). Pure Sertoli cell cultures: A new model for the study of somatic-germ cell interactions. *J. Androl.* **2,** 249–254.

Galdieri, M., Monaco, L., and Stefanini, M. (1984a). Secretion of androgen-binding protein by Sertoli cells is influenced by contact with germ cells. *J. Androl.* **5,** 409–415.

Galdieri, M., Monaco, L., and De Martino, C. (1984b). Morphological and biochemical effects of Lonidamine on cultured Sertoli cells. *Exp. Mol. Pathol.* **41,** 202–206.

Golberg, L. (1986). Charting a course for cell culture alternatives to animal testing. *Fundam. Appl. Toxicol.* **6,** 607–617.

Gray, T. J. B., and Beamand, J. A. (1984). Effect of some phthalate esters and other testicular toxins on primary cultures of testicular cells. *Food Chem. Toxicol.* **22,** 123–131.

Gray, T. J. B., Rowland, I. R., Foster, P. M. D., and Gangolli, S. D. (1982). Species differences in the testicular of phthalate esters. *Toxicol. Lett.* **11,** 141–147.

Gray, T. J. B., Moss, E. J., Creasy, D. M., and Gangolli, S. D. (1985). Studies on the toxicity of some glycol ethers and alkoxyacetic acids in primary testicular cell cultures. *Toxicol. Appl. Pharmacol.* **79,** 490–501.

Grisham, J.W., and Smith, G. J. (1984). Predictive and mechanistic evaluation of toxic responses in mammalian cell culture systems. *Pharmacol. Rev.* **36,** 1515–1715.

Grootegoed, J. A., Krüger-Sewnarain, B. C., Jutte, N. H. P. M., Rommerts, F. F. G., and van der Molen, H. J. (1982). Fucosylation of glycoproteins in rat spermatocytes and spermatids. *Gamete Res.* **5,** 303–315.

Hadley, M. A., Byers, S. W., Suárez-Quian, C. A., Kleinman, H. K., and Dym, M. (1985). Extracellular matrix regulates Sertoli cell differentiation, testicular cord formation, and germ cell development *in vitro. J. Cell Biol.* **101,** 1511–1522.

Hagenäs, L., Plöen, L., and Ritzén, E. M. (1978). The effect of nitrofurazone on the endocrine, secretory and spermatogenic functions of the rat testis. *Andrologia* **10,** 107–126.

Hall, P. F., and Mita, M. (1984). Influence of follicle-stimulating hormone on glucose transport by cultured Sertoli cells. *Biol. Reprod.* **31,** 863–869.

Janecki, A., Lukaszyk, A., and Sobieszczyk, S. (1984). Stimulatory effect of Sertoli cell culture medium on the FSH release by pituitary halves *in vitro. Horm. Metab. Res.* **16,** 581–584.

Jutte, N. H. P. M., Jansen, R., Grootegoed, J. A., Rommerts, F. F. G., and van der Molen, H. J. (1983). FSH stimulation of the production of pyruvate and lactate by rat Sertoli cells may be involved in hormonal regulation of spermatogenesis. *J. Reprod. Fertil.* **68,** 219–226.

Kierszenbaum, A. L., and Tres, L. L. (1981). The structural and functional cycle of Sertoli cells in culture. *In* "Bioregulators of Reproduction" (G. Jagiello and H. J. Vogel, eds.), pp. 207–228. Academic Press, New York.

Kierszenbaum, A. L., DePhillip, R.M., Spruill, W. A., and Takenaka, I. (1983). Isolation and culture of rat seminal vesicle epithelial cells. *Exp. Cell Res.* **145,** 293–304.

Laskey, J. W., Phelps, P. V., and Laws, S. D. (1986). Leydig cell function *in vitro* following metal cation treatment. *Toxicologist* **6,** 86, abstr. 344.

Laws, A. O., Wreford, N. G. M., and de Kretser, D. M. (1985). Morphological and functional characteristics of rat Leydig cells isolated on Percoll gradients: is Leydig cell heterogeneity *in vitro* an artefact? *Mol. Cell. Endocrinol.* **42,** 73–90.

Lee, I. P., and Nagayama, J. (1980). Metabolism of benzo(a)pyrene by the isolated perfused rat testis. *Cancer Res.* **40,** 3297–3303.

Mather, J. P., and Phillips, D. M. (1984). Primary culture of testicular somatic cells. *In* "Cell Culture Methods for Molecular and Cell Biology. Vol 2. Methods for Serum-Free Culture of Cells of the Endocrine System" (D. W. Barnes, D. A. Sirbasku, and G. H. Sato, eds.), pp. 29–45. Alan R. Liss, New York.

Mather, J. P., Perez-Infante, L. Z., Zhuang, V., and Phillips, D.M. (1982). Culture of testicular cells in hormone-supplemented serum-free medium. *Ann N.Y. Acad. Sci.* **383,** 44–68.

Meistrich, M. L. (1977). Separation of spermatogenic cells and nuclei from rodent testes. *In* "Methods in Cell Biology," Vol. 15 (D. M. Prescott, ed.), pp. 15–54. Academic Press, New York.

Meistrich, M. L., Longtin, J., Brock, W. A., Grimes, S. R., and Mace, M. L. (1981). Purification of rat spermatogenic cells and preliminary biochemical analysis of these cells. *Biol. Reprod.* **25,** 1065–1077.

Miller, G. E., Beattie, P. J., Welsh, M. J., and Brabec, M. J. (1986). Comparative cytotoxicity of 1,2-dibromo-3-chloropropane (DBCP) and proposed metabolites to isolated spermatocytes. *Toxicologist* **6,** 289, abstr. 1157.

Molenaar, R., Rommerts, F. F. G., and van der Molen, H. J. (1983). The steroidogenic activity of isolated Leydig cells from mature rats depends on the isolation procedure. *Int. J. Androl.* **6,** 261–274.

Molenaar, R., de Rooij, D. G., Rommerts, F. F. G., Reuvers, P. J., and van der Molen, H. J. (1985). Specific destruction of Leydig cells in mature rats after *in vivo* administration of ethane dimethyl sulfonate. *Biol. Reprod.* **33,** 1213–1222.

Moore, H. D. M. (1986). *In vitro* manipulation of gametes and embryos for evaluating the effect of chemicals on germ cell function and developmental potential. *Food Chem. Toxicol.* *24*, 607–613.

Moss, E. J., Cook, W. M., Thomas, L. V., Savage, A. C. T. Creasy, D. M., and Gray, T. J. B. (1986). Effect of mono-(2-ethylhexyl) phthalate and other phthalate monoesters on lactate production by Sertoli cell-enriched cultures. *Human Toxicol.* **5** 144–145.

O'Connor, T., and Sinha, D. K. (1985). Characterisation of rat ventral prostatic epithelial cells in collagen gel culture, *Prostate* **7**, 303–319.

Oudiz, D., and Zenick, H. (1986). *In vivo* and *in vitro* evaluations of spermatotoxicity induced by 2-ethoxyethanol treatment. *Toxicol. Appl. Pharmacol.* **84,** 576–583.

Perrard-Sapori, M. H., Chatelain, P.,Vallier, P., and Saez, J. M. (1986). *In vitro* interactions between Sertoli cells and steroidogenic cells. *Biochem. Biophys. Res. Commun.* **134,** 957–962.

Rich, K. A., Bardin, C. W., Gunsalus, G. L., and Mather, J. P. (1983). Age-dependent pattern of androgen-binding protein secretion from rat Sertoli cells in primary culture. *Endocrinology* **113,** 2284–2293.

Rommerts, F. F. G., Molenaar, R., and van der Molen, H. J. (1985). Preparation of isolated Leydig cells. *In* "Methods in Enzymology," Vol. 109 (L. Birnbaumer and B. W. O'Malley, eds.), pp. 275–288. Academic Press, New York.

Romrell, L. J., Bellvé, A. R., and Fawcett, D. W. (1976). Separation of mouse spermatogenic cells by sedimentation velocity. A morphological characterization. *Dev. Biol.* **49,** 119–131.

Rowan, A. N., and Goldberg, A. M. (1985). Perspectives on alternatives to current animal testing techniques in preclinical toxicology. *Annu. Rev. Pharmacol. Toxicol.* **25,** 225–247.

Skinner, M. K., and Fritz, I. B. (1986). Identification of a non-mitogenic paracrine factor involved in mesenchymal-epithelial cell interactions between testicular peritubular cells and Sertoli cells. *Mol. Cell Endocrinol.* **44,** 85–97.

Spitz, I. M., Gunsalus, G. L., Mather, J. P., Thau, R., and Bardin, C. W. (1985). The effects of the indazole carboxylic acid derivative, Tolnidamine, on testicular function: I. Early changes in androgen binding protein secretion in the rat. *J. Androl.* **6,** 171–178.

Steinberger, A., Heindel, J. J., Lindsey, J. N., Elkington, J. S. H., Sanborn, B. M., and Steinberger, E. (1975). Isolation and culture of FSH responsive Sertoli cells. *Endocr. Res. Commun.* **2,** 261–272.

Stern, L., Gold, B, and Hecht, N. B. (1983). Gene expression during mammalian spermatogenesis. 1. Evidence for stage-specific synthesis of polypeptides *in vivo*. *Biol. Reprod.* **28,** 483–496.

Syed, V., Khan, S. A., and Ritzen, E. M. (1985). Stage-specific inhibition of interstitial cell testosterone secretion by rat seminiferous tubules *in vitro*. *Mol. Cell. Endocrinol.* **40,** 257–264.

Tanphaichitr, N., Chen, L. B., and Bellvé, A. R. (1984). Direct effect of Gossypol on TR-ST cells: perturbation of Rhodamine 123 accumulation in mitochondria. *Biol. Reprod.* **31,** 1049–1060.

Toppari, J., and Parvinen, M. (1985). *In vitro* differentiation of rat seminiferous tubular segments from defined stages of the epithelial cycle: Morphologic and immunolocalization analysis. *J. Androl.* **6,** 334–343.

Tres, L. L., and Kierszenbaum, A. L. (1983). Viability of rat spermatogenic cells *in vitro* is facilitated by their coculture with Sertoli cells in serum-free hormone-supplemented medium. *Proc. Natl. Acad. Sci. U.S.A.* **80,** 3377–3381.

Wagle, J. R., Heindel, J. J., Steinberger, A., and Sanborn, B. M. (1986). Effect of hypotonic treatment on Sertoli cell purity and function in culture. *In Vitro* **22,** 325–331.

Welsh, M. J., and Weibe, J. P. (1975). Rat Sertoli cells: A rapid method for obtaining viable cells. *Endocrinology* **96,** 618–624.

Yanagimachi, R., Yanagimachi, H., and Rogers, B. J. (1976). The use of zona-free animal ova as a test system for the assessment of the fertilizing capacity of human spermatozoa. *Biol. Reprod.* **15,** 471–476.

Yee, J. B., and Hutson, J. C. (1985a). Biochemical consequences of follicle-stimulating hormone binding to testicular macrophages in culture. *Biol. Reprod.* **32,** 872–879.

Yee, J. B., and Hutson, J. C. (1985b). Effects of testicular macrophage-conditioned medium on Leydig cells in culture. *Endocrinology,* **116,** 2682–2684.

Zhuang, L.-Z., Phillips, D. M., Gunsalus, G. L., Bardin, C. W., and Mather, J. P. (1983). Effects of Gossypol on rat Sertoli and Leydig cells in primary culture and established cell lines. *J. Androl.* **4,** 336–344.

Index

Q

R

T

U

V

W

X

Y

Z